SOCIO-HYDROLOGICAL DYNAMICS IN BANGLADESH

Understanding the interaction between hydrological and social processes

along the Jamuna floodplain

Md Ruknul Ferdous

SOCIO-HYDROLOGICAL DYNAMICS IN BANGLADESH

Understanding the interaction between hydrological and social processes along the
Jamuna floodplain

ACADEMISCH PROEFSCHRIFT

ter verkrijging van de graad van doctor
aan de Universiteit van Amsterdam
op gezag van de Rector Magnificus
prof. dr. ir. K.I.J. Maex
ten overstaan van een door het College voor Promoties ingestelde commissie,
in het openbaar te verdedigen in de Agnietenkapel
op woensdag 15 januari 2020, te 10:00 uur

door

Md Ruknul Ferdous

geboren te Dinajpur

Promotiecommissie:

Promotores:	Prof. dr. M.Z. Zwarteveen	Universiteit van Amsterdam
	Prof. dr. G. Di Baldassarre	Uppsala universitet
Copromotores:	Dr. L. Brandimarte	Kungliga Tekniska högskolan
	Dr. A.J. Wesselink	IHE Delft
Overige leden:	Prof. dr. J. Gupta	Universiteit van Amsterdam
	Prof. dr. J.C.J.H. Aerts	Vrije Universiteit Amsterdam
	Prof. dr. P. van der Zaag	IHE Delft / TU Delft
	Prof. dr. P.J. Ward	Vrije Universiteit Amsterdam
	Dr. H. Kreibich	GFZ Potsdam

Faculteit der Maatschappij- en Gedragswetenschappen

To my parents Rezina Khatoon and late Alauddin Ahmed

Published by:
CRC Press/Balkema
Schipholweg 107C, 2316 XC, Leiden, the Netherlands
Pub.NL@taylorandfrancis.com
www.crcpress.com – www.taylorandfrancis.com
ISBN 978-0-367-90213-1 (Taylor & Francis Group)

ACKNOWLEDGMENTS

I am extremely grateful to Almighty for enabling me to complete my PhD research. I would like to acknowledge and thank many people for their continuous help and support while conducting my PhD research.

First of all, I would like to express my gratitude and respect to my supervisor and promotor, Professor Margreet Zwarteveen, for giving me the opportunity to work as a PhD student under her supervision. I am extremely grateful to Margreet for her contribution from the beginning to the end of my PhD research. I learned a lot from Margreet throughout the period as a PhD student. I thank her for her outstanding supervision through her scientific knowledge and interest.

I would also like to express my gratitude and respect to my supervisor and co-promotor, Professor Giuliano Di Baldassarre, for being with me during the entire period of my PhD research. He also supervised my MSc thesis and this was the reason I undertook the PhD. I am highly thankful and grateful to Giuliano for his support, guidance, enthusiasm and valuable input in my PhD research. Without his kind support and guidance it would have been impossible for me to finish my PhD.

I would like to express my sincere gratitude and heartfelt thanks to my co-supervisor, Dr. Luigia Brandimarte for her interest and careful supervision and guidance right from the beginning of my PhD research. She also supervised my MSc thesis and this was also the reason I undertook the PhD. She helped me a lot in every aspect of my PhD research and I learned a lot from Luigia throughout the period as a PhD student. This PhD thesis would have been impossible without her unconditional inputs.

I would like to express my heartfelt gratitude and thanks to my co-supervisor, Dr. Anna Wesselink, for her interest and continuous support from the beginning of my PhD research. Without her kind support, guidance, enthusiasm to help and extreme tolerance it would have been impossible for me to finish my PhD. I am grateful and thankful from bottom of my heart to Anna for her immense contribution for my PhD research. I learned a lot from Anna throughout the period of my research as a PhD student. She was always beside me during the entire period of my PhD research.

I would like to express my sincere gratitude and thanks to Dr. Michelle Kooy and Professor Arthur Mynett for their interest and careful guidance at the beginning of my PhD research. This PhD thesis would not have been possible without their unconditional inputs during my PhD proposal development.

I am also thankful to Dr. Kymo Slager of Deltares for his cordial co-operation and perceptive guidance for my PhD research. I express my heartfelt gratitude to him for his unconditional help to write two scientific journal papers. I am really grateful to him.

I am thankful to Dr. Parvin Sultana and Dr. Paul Thompson of FHRC Bangladesh for their cordial co-operation and guidance in developing my PhD proposal. I am also grateful to them for helping me during field work at Bangladesh.

I express my heartfelt gratitude to Mr. Md Mahabubar Rahman, Mr. Md Enamul Haque and Mr. Md Shahrier Islam of Gaibandha, Bangladesh, who were my field assistants in Bangladesh. Without their support it would not have been possible for me to collect household survey data. I am really grateful to them.

I express my sincere gratitude to CEGIS for providing hydrological data and satellite images for my PhD research. I am really grateful to CEGIS for their help and support.

I am thankful to Mr. Sudipta Kumar Hore, Ms. Deeba Farzana Moumita and Mr. Mohammad Saidur Rahman for their help and support to do satellite image analysis during my PhD research. I am also thankful to Mr. Md Mizanur Rahman for allowing me to use his MSc thesis data in my PhD research. I am really grateful to them.

I am grateful to NWO-WOTRO for funding my PhD research through the project "Hydro-Social Deltas: Understanding flows of water and people to improve policies and strategies for disaster risk reduction and sustainable development of delta areas in the Netherlands and Bangladesh". I also thank IHE Delft Institute for Water Education, the Netherlands and University of Amsterdam, the Netherlands for hosting me during my PhD research period.

I am also grateful to all of my ex-colleagues and friends in Bangladesh to help and support me during my PhD research. I am also thankful to all my colleagues and friends here in the Netherlands for entertaining me during my stay in the Netherlands.

I would like to convey my immense love and gratitude to my beloved mother Rezina Khatoon and my loving sisters Jannatul Ferdous and Bentul Mawa, my brother Zannatul Adan, for their endless love, encouragement and inspiration without which I find everything incomplete and barren. I also like to remember my late beloved father Alauddin Ahmed for his endless love and affection for me. He always encouraged me and showed me the right path.

Most importantly I thank from the inner core of my heart my beloved wife Shamrin Rahman, who has taken all the burdens of my family during my entire PhD period here in the Netherlands. She took care of our son and helped me to complete my work. It would not have been possible for me to complete my PhD without her endless love, unconditional support, patience, and blessings. I feel deeply indebted to her. I am also blessed to have my son Sheaaf Ibrahim Ferdous with me here in the Netherlands. He is always an inspiration to me. My tiredness vanishes with his smile and boosts energy for my work.

SUMMARY

Bangladesh is a large delta, where most people live in the overpopulated floodplains. Flooding and riverbank erosion are normal phenomena, which cause lots of suffering and demographic shifts, livelihood changes, impoverishment, etc. Society takes different initiatives such as the construction of embankments along the river bank, aiming to control hydrological processes. The study of interactions and feedback mechanisms between hydrological and social processes is a new academic field, particularly relevant in a dynamic delta such as Bangladesh. Socio-hydrological research in floodplains aims to unravel the two-way interactions between floodplains societies and water systems. As socio-hydrology is related to short-term and long-term dynamics, understanding possible system dynamics for the future would be of interest to governments who are dealing with strategic and long term-decisions.

The principal objective of this research is to better understand the interactions between physical processes and societal processes along the Jamuna River in Bangladesh. They are conceptualised as temporally dynamic and spatially diverse combinations of 'fighting with water' and 'living with water'. The objectives of this research are achieved by means of spatial and temporal analysis of social and physical data on a short stretch of the Jamuna floodplain near Gaibandha and Jamalpur of Bangladesh. Primary data were collected through 900 household surveys and 15 focus group discussions. Secondary data (river water level, erosion, population, satellite images) were collected from governmental agencies. The dynamics of hydrological processes and social processes have been analysed by statistical analyses and mapping with the aid of GIS.

This PhD research proposes a concept "socio-hydrological spaces (SHSs)" that enriches the study of socio-hydrology because it helps understand the detailed human-water interactions in a specific location. A socio-hydrological space is a geographical area in a landscape. Its particular combination of hydrological and social features gives rise to the emergence of distinct interactions and dynamics (patterns) between society and water. The SHSs concept suggests that the interactions between society and water are place-bound and specific because of differences in social processes, technological choices and opportunities, and hydrological dynamics. Such attention is useful anywhere in the world and also for other socio-hydrological systems than floodplains. To apply this concept in-depth knowledge of a place is very important. The in-depth knowledge or past experience will help a researcher to identify the socio-hydrological spaces of a specific location. In terms of practical use, it can be added as an additional element to rapid rural appraisals, or other social assessments, to draw attention to how material conditions (hydrological and technical/infrastructure) co-shape social situations. I applied this concept in my study area at Jamuna floodplain and acquired better understanding on human-water interaction on that area.

This PhD research applied the SHSs concept while analysing another objective, the costs of living with floods in the Jamuna floodplain. It is observed that the respondents of the whole study area are falling in impoverishment in the long run but the costs and experiences of living with floods are different with different socio-hydrological spaces. I also explored the 'levee effect' in three districts along the Jamuna River by comparing the historical socio-economic development and the current state of affairs. The 'levee effect' was first proposed by White (1945). He asserted that the construction of a levee to protect from flooding might induce property owners to invest more in their property, increasing the potential damages should the levee breach. Paradoxically, flood risk may thereby rise as a consequence. The Jamuna floodplain presents different flood protection measures on the two river banks. The Brahmaputra Right Embankment (BRE), a man-made earthen levee, was built in the 1960s parallel to the right bank, but no similar structural investment was made on the left bank, leaving it unprotected. Furthermore, at the present time the BRE shows different characteristics along its length, with stronger levees (reinforced with concrete, and maintained) in the southern part of the right bank, and weaker (unmaintained) levees in the northern part. These differences enable me to compare the effect of different protection levels within an otherwise homogeneous region. The results support consolidated theories about the interplay between levels of structural flood protection, people and assets exposed to flooding, and social vulnerability to flooding.

This PhD research does not only contribute to advance the knowledge about socio-hydrological dynamics in Bangladesh, but also provides more general insights for flood risk management. A better understanding of the dynamics between the hydrological and social processes in Jamuna floodplain can inform policy making aimed at reducing risk and vulnerability and to select the best adaptation or mitigation options for Bangladesh.

SAMENVATTING

Bangladesh is een grote delta, waar de meeste mensen in de overbevolkte uiterwaarden wonen. Overstromingen en rivieroeversosie zijn normale verschijnselen, die veel lijden en demografische verschuivingen, verandering van levensonderhoud, verarming, enz. veroorzaken. De maatschappij neemt verschillende initiatieven, zoals de aanleg van dijken, gericht op het beheersen van hydrologische processen. De studie van interacties en feedbackmechanismen tussen hydrologische en sociale processen is een nieuw academisch veld, met name relevant in een dynamische delta zoals Bangladesh. Sociaal-hydrologisch onderzoek in uiterwaarden heeft tot doel de wederzijdse interacties tussen de samenleving in uiterwaarden en de watersystemen te ontrafelen. Aangezien socio-hydrologie gerelateerd is aan de dynamiek op korte en lange termijn, is het begrijpen van deze systeemdynamica mogelijk van belang voor overheden die te maken hebben met strategische en langetermijn beslissingen.

Het belangrijkste doel van dit onderzoek is daarmee om de interacties tussen fysieke processen en maatschappelijke processen langs de Jamuna rivier in Bangladesh beter te begrijpen. Deze interactie worden geconceptualiseerd als temporeel dynamische en ruimtelijk diverse combinaties van 'vechten met water' en 'leven met water'. De doelstellingen van dit onderzoek worden bereikt door middel van ruimtelijke en temporele analyse van sociale en fysieke gegevens op een kort stuk van de Jamuna-uiterwaarden nabij Gaibandha en Jamalpur van Bangladesh. Primaire gegevens werden verzameld via 900 huishoudensquêtes en 15 focusgroep discussies. Secundaire gegevens (rivierwaterstand, erosie, bevolking, satellietbeelden) werden verzameld bij overheidsinstanties. De dynamiek van hydrologische processen en sociale processen is geanalyseerd door middel van statistische analyses en met behulp van GIS.

Dit doctoraatsonderzoek stelt ook een nieuw begrip voor "socio-hydrological spaces (SHSs)" dat de studie van socio-hydrologie verrijkt omdat het helpt de gedetailleerde mens-water interacties op een specifieke locatie te begrijpen. Een SHS is een geografisch gebied in een landschap. De specifieke combinatie van hydrologische en sociale kenmerken geeft aanleiding tot de opkomst van verschillende interacties en dynamieken (patronen) tussen samenleving en water. Het SHS-concept suggereert dat de interacties tussen samenleving en water plaatsgebonden en specifiek zijn vanwege verschillen in sociale processen, technologische keuzes en kansen en hydrologische dynamiek. Dergelijke aandacht is overal ter wereld nuttig en ook voor andere sociaal-hydrologische systemen dan uiterwaarden. Om dit concept toe te passen is een grondige kennis van een plaats erg belangrijk. De diepgaande kennis of ervaring uit het verleden zal een onderzoeker helpen om de SHS van een specifieke locatie te identificeren. In termen van praktisch gebruik kan het als een extra element worden toegevoegd aan snelle beoordelingen op het platteland, of andere sociale beoordelingen, om de aandacht te vestigen op hoe materiële omstandigheden (hydrologisch en technisch / infrastructuur) samenhangen met sociale omstandigheden en processen. Ik heb dit concept toegepast in mijn studiegebied en daarmee heb ik een beter begrip gekregen van de wisselwerking tussen mens en water in het gebied.

Dit promotieonderzoek paste het SHS-concept daarnaast toe bij het analyseren van een andere doelstelling, de kosten van het leven met overstromingen in de Jamuna-uiterwaard. Opgemerkt wordt dat de respondenten van het hele studiegebied op de lange termijn in verarmen, maar de ervaringen en consequenties van het leven met overstromingen verschillen in de verschillende SHS. Ik heb ook het 'dijkeffect' in drie districten langs de Jamuna-rivier verkend door de historische sociaal-economische ontwikkeling en de huidige stand van zaken te vergelijken. Het 'dijkeffect' werd voor het eerst voorgesteld door White (1945). Hij beweerde dat de bouw van een dijk om te beschermen tegen overstromingen ertoe zou kunnen leiden dat eigenaren van gebouwen meer in hun eigendom investeren, waardoor de potentiële schade toeneemt als de dijk doorbreekt. Paradoxaal genoeg kan daardoor het overstromingsrisico toenemen. In de uiterwaarden van de Jamuna zijn verschillende maatregelen ter bescherming tegen overstromingen geconstrueerd op de twee rivieroevers. De Brahmaputra Right Embankment (BRE), aarden dijk, werd gebouwd in de jaren 60 op de rechteroever, maar er werd geen vergelijkbare structurele investering gedaan op de linkeroever, waardoor deze onbeschermd bleef. Bovendien vertoont het BRE op dit moment verschillende kenmerken afhankelijk van de locatie, met sterkere dijken (versterkt met beton en goed onderhouden) in het zuidelijk deel van de rechteroever en zwakkere (niet-onderhouden) dijken in het noordelijke deel. Deze verschillen stellen mij in staat om het effect van verschillende beschermingsniveaus binnen een anderszins homogene regio te vergelijken. Mijn resultaten ondersteunen de theorieën over de wisselwerking tussen structurele bescherming, mensen, kapitaal en kwetsbaarheid voor overstromingen.

Dit promotieonderzoek draagt niet alleen bij aan het vergroten van de kennis over de sociaal-hydrologische dynamiek in Bangladesh, maar biedt ook meer algemene inzichten voor overstromingsrisicobeheer. Een beter begrip van de dynamiek tussen de hydrologische en sociale processen in de overstromingsvlakte van Jamuna kan helpen bij het maken van beleid gericht op het verminderen van risico's en kwetsbaarheid en het selecteren van de beste adaptatie- of mitigatieopties voor Bangladesh.

TABLE OF CONTENTS

Table of contents
___

1

INTRODUCTION

1.1 BACKGROUND

Water is essential to life and rivers are one of the main sources of water. Civilizations such as those in China, Egypt, Babylon and India have developed along and thanks to famous rivers: the Yellow, Nile, Euphrates and Tigris and Ganges respectively (Di Baldassarre et al., 2013a; Liu et al., 2014). This is because the floodplains of these rivers constitute very favourable areas in terms of geographical location and environmental conditions to settle in, even though they are also prone to flooding (Junk et al., 1989). Today still, floodplains remain attractive places to live and settle, places where societies can practice agriculture and find means of transportation for trade and economic growth (Di Baldassarre et al., 2013a). Yet, living in floodplains comes with risks. Flooding and the relocation of river channels by erosion and deposition may cause damage to agricultural lands, crops, homes, roads and buildings. When deciding whether to live in a floodplain area, people therefore evaluate the chances of this damage to occur, including the extent of flooding they may experience, and make a decision whether to mitigate the risks or live with them.

As the number of people living in floodplains continues to increase (Lutz, 1997), so do the risks societies living in floodplains face. Flood risk can be defined as a combination of the probability of a flood event and its potential adverse consequences, or risk = probability * consequences (Di Baldassarre et al., 2009; van Manen and Brinkhuis, 2005). Vulnerability is the degree to which societies are susceptible to destruction by hostile factors (Şorcaru, 2013) and includes societies' responses to risk as well as the risks themselves. Flood risks are not usually equally distributed across societies, with some people having a better ability to cope than others (Masozera et al., 2007; Di Baldassarre et al., 2013a,b). Societal responses to hazards in floodplains can include demographic shifts (such as resettlements and migration), governance measures (such as flood control, disaster management), land use changes, livelihood changes (such as a change in occupation), and technical interventions in the physical layout of the floodplain, e.g. homestead mounds, ditches and dikes. Such responses may in turn influence the future behaviour of rivers and the chances of floods; they may for instance exacerbate flood risk rather than reducing it, thus increasing the vulnerability of a society (Milly et al., 2002; Di Baldassarre et al., 2010).

Such dynamic interactions between society and rivers in floodplain areas have become a topic of research interest (Di Baldassarre et al., 2013a,b). Studies for instance look at how constructing flood control measures or changing land-use patterns alter the frequency and severity of floods. As Di Baldassarre et al., (2013a,b) argue, however, the dynamic interactions and associated feedback mechanisms between hydrological and social processes remain unexplored and poorly understood. Indeed, there is a lack of understanding on how and to what extent hydrological processes influence or trigger changes in social processes and vice-versa. To address this gap in knowledge, Sivapalan et al., (2012) have proposed the new science of socio-hydrology. Socio-hydrology proposes to study the two-way coupling of human and water systems with the aim of advancing the science of hydrology for the benefit of society (Montanari et al., 2013). This research takes inspiration from and aims to advance socio-hydrology. It focusses on the two-way interactions between society and water in the Jamuna floodplain in Bangladesh.

1.2 SOCIETAL RELEVANCE AND SELECTION OF CASE STUDY

More than 100 million people in the world are affected by floods every year. Flooding is the most damaging of all so-called natural hazards, causing about fifty percent of all deaths from disasters worldwide (Ohl and Tapsell, 2000; Opperman et al., 2009). Increases in population and demographic expansion enhance pressures on river floodplains, resulting in increased flood probability and hence risk. In the years 2011 and 2012 alone, about 200 million people were affected by floods with a total damage about 95 billion USD 2012 (Ceola et al., 2014).

Bangladesh is among the countries that is notoriously prone to flooding because of its location at the confluence of three mighty rivers: the Ganges, Brahmaputra and Meghna. Indeed, Bangladesh is one of the largest deltas in the world (Mirza et al., 2003). The combined flows of the three rivers come together in the Bay of Bengal through the Lower Meghna River, culminating in a total of 1 trillion m^3/year of water and 1 billion tonnes/year of sediment (Allison, 1998). In addition, there are hundreds of medium and small tributaries and distributaries flowing through the country. More than half of the country consists of floodplain areas (Tingsanchali et al., 2005; Nardi et al., 2019). About 10% of the total area of the country is less than 1 m above mean sea level and one-third is under tidal influence (Ali, 1999). 92.5% of the combined basin area of Ganges, Brahmaputra and Meghna lies outside of the country, putting Bangladesh at the mercy of upstream water uses and management decisions. With 80% of the annual rainfall occurring in the monsoon (June-September) across these river basins (Mirza, 2002), runoff is concentrated in just a few months. Very often, the peak volume exceeds the capacity of the river channel networks, resulting in severe flood events. Such events happened in 1954, 1955, 1974, 1987, 1988, 1998, 2004, 2007 and 2017 (FFWC/BWDB, 2018). The most severe flood events occurred in the last 50 years. In 1988 and 1998, about 61% and 68% respectively of the country was inundated by riverine flooding (Penning-Rowsell et. al, 2012). While the northern part of Bangladesh is vulnerable to riverine flooding only, the southern coastal part is mainly vulnerable to flooding due to storm surges (Tingsanchali et al. 2005). There are several other factors which make Bangladesh prone to flooding: very low floodplain gradients, the reduction of stream channel capacities due to the building of infrastructures such as dykes and polders, and excessive siltation of river beds (Rahman, 1996). In addition, Bangladesh is experiencing subsoil compaction and iso-statically induced subsidence, which further increase flooding (Goodbred and Kuehl, 1999, Sarker et al. 2003).

Bangladesh is a very densely populated country with more than 140 million of people (964 persons per km^2). Around 80% of the total population lives in floodplain areas and most of them mainly depends on agriculture (BBS, 2013). Flooding and riverbank erosion severely impact agriculture through the loss of land and crops. The 1988 flood affected about 73 million people (Penning-Rowsell et. al, 2012), among them about 45 million people were displaced and about 2,000-6,500 died. The 1998 floods caused about 1,100 deaths and displaced about 30 million people (IOM, 2010). In the most recent severe floods of 2017, about 62,000 km^2 (about 42% area of the country) was inundated (FFWC/BWDB, 2018), affecting about 8.3 million people and causing 147 deaths

(NDRCC, 2017). Around 700,000 houses were severely damaged and 0.7 million hectares of cropland affected (NDRCC, 2017).

River channels and flooding patterns are changing every year. The extremely poor people who are living on the so-called chars - islands in the big rivers - are most exposed to and affected by flood hazards and riverbank erosion. During severe flood events, they often have to migrate to nearby areas. They may have to move permanently because the river islands they lived and farmed on simply disappear. During 1981-1993, about 0.7 million char-land dwellers were displaced. Half of them were from chars in the Jamuna River (FAP 16/19 1993).

Successive governments have developed a range of measures to protect agriculture and populations from floods (Sultana et al., 2008). However, the country encounters and has encountered difficulties to mobilize enough funds to adequately protect all areas. This is why the government decided to prioritize some areas to be protected first. After the severe floods in the years 1954 and 1955, the government for instance decided to construct an embankment along the west bank of the Jamuna River, called the "Brahmaputra Right Embankment (BRE)". The main reason for this choice was that the duration of the flood was longer on the west bank than on the east bank, something that is caused by the average land elevation being lower along the west bank than along the east bank (Sarker et al., 2014). In addition, crop production was higher along the west bank, which is why total damage on the west bank was also higher. In addition, the river has been eroding and migrating westward since 1830 (Coleman, 1969). The BRE was constructed in the 1960s to limit flooding in an area of about 240,000 ha lying on the western and southern sides of the Brahmaputra-Jamuna and Teesta rivers respectively and to increase the agricultural production of that area. The construction of the embankment started in 1963 and was completed in 1968 at a cost of about BDT 80 million (~16.7 million USD) (CEGIS, 2007). This was the first major intervention to protect people and lands from flooding in Bangladesh. In addition to such physical interventions, the government also set up a programme to predict flooding since 1972 (FFWC/BWDB, 2018) and riverbank erosion since 2004 (CEGIS, 2016) and disseminate the results to the affected population, to increase their ability to prepare for and cope with flooding and riverbank erosion.

5

A scoping study has indicated that the impacts of embankments are not necessarily and always positive (Burton and Cutter, 2008; Opperman et al., 2009). To date, no studies have been done to better understand the relation between social and hydrological processes along the main floodplain areas of Bangladesh. The few studies that do exist aimed to understand these relations in the coastal region of Bangladesh. These have shown how human interventions such as upstream diversion and coastal polders significantly contributed to the variation in water salinity, tidal water level and flooding in the southwest coastal region of Bangladesh (Mondal et al., 2013, Ferdous, 2014). Other studies (e.g. Walsham, 2010) show that hydrological changes in turn shape patterns of human settlements and trigger migration. However, these studies always look at one or the other side of the interplay between hydrological and social processes. The dynamic interactions and feedbacks between social processes and hydrological processes remain underexplored. Studying these is a new academic field, one that is particularly relevant in a dynamic delta such as Bangladesh. The purpose of the present study precisely is to delve into these interactions and feedbacks, with the purpose to better understand the dynamic relations between social and hydrological processes in the main floodplain areas of Bangladesh. The ultimate aim of producing a better understanding is to help developing better protection systems, and reduce the damage and suffering caused by flooding and riverbank erosion. The hope is that improved understanding of these interactions between social and hydrological processes using a coupled two-way dynamic conceptual model will strengthen our capability to design suitable future interventions and/or suitable adaptation options in Bangladesh.

The research was carried out in a case study area that includes rural areas in Gaibandha and Jamalpur districts and two urban areas Gaibandha town and Sirajganj town (Figure 1.1). The total surface of the study area is about 550 km^2 and the total population is approx. 0.6 million (BBS, 2013, BBS, 2014a). I have chosen it as my case study area because of my familiarity with it: I have more than 10 years of professional experience working here as a water engineer charged with flood forecasting, flood management and training residents on using flood forecasts. During my professional work, I observed how difficult it is to take decisions to manage floods in this region. Hence, I am directly motivated to and interested in helping decision makers to do flood management and take flood management decisions. More detailed descriptions of the case study areas are presented in chapters 2, 3, 4 and 5.

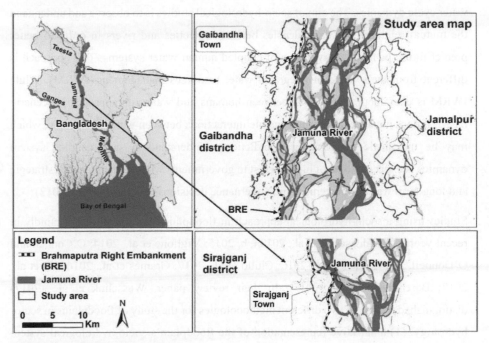

Figure 1.1: Map of the study area.

1.3 CONCEPTS AND THEORIES

Water resource studies in floodplains were until recently dominated by hydrologic studies for water resources planning, flood management etc. aimed at informing engineering interventions. In recent years, since the 1990s, hydrologic studies have started to integrate social variables and data in their analysis (McKinney, 1999). This was in recognition of the fact that ignoring 'the human factor' in dealing with and making decisions about water and rivers was dangerous (Sivapalan et al., 2012). Nowadays hydrologists are gradually entering into and developing the field of socio hydrology, which outs out to advance understanding on how and to what extent hydrological processes influence or trigger changes in social processes and vice-versa. In traditional hydrology, human factors are assumed as external forces or taken as the given, stationary context in which water cycle dynamics play out (Milly et al., 2008; Peel and Blöschl, 2011). Socio-hydrology instead aspires to include humans and societal dynamics and processes as internally related to water cycle dynamics. The aim of socio-hydrological studies is to observe and understand the mutual relations and dependencies between societies and rivers in order to better predict future paths of co-evolution of coupled human-water systems. This approach is different from the science of integrated water resources management (IWRM). While IWRM is also about interactions between humans and water, its approach is scenario based. It does not account for the dynamic interactions between water and people, which may be unrealistic for long-term predictions. Understanding such coupled system dynamics is expected to be of high interest to governments who are dealing with strategic and long term water management and governance decisions (Sivapalan et al., 2012).

Studies using a socio-hydrological approach in floodplains have proliferated rapidly in recent years (Di Baldassarre et al., 2013a, b, 2015; Viglione et al., 2014; O'Connell and O'Donnell, 2014; Chen et al., 2016; Ciullo et al., 2017; Grames et al., 2016; Yu et al., 2017; Barendrecht et al., 2017). In their review paper, Wesselink et al., (2017) distinguished the use of two different methodologies for the study of floodplains in socio-hydrology: (1) a narrative representation of the floodplain's socio-hydrological system based on qualitative research and (2) a characterisation of the development of the floodplain through the use of a generic conceptual model of human–nature interactions which is expressed in terms of differential equations. The first approach is importantly led by experiences and knowledge of local inhabitants and experts about their histories of living with floods. Sometimes it may also include historical statistical data to see trends. These studies are useful for capturing historical patterns in the co-evolution of river dynamics, settlement patterns and technological choices (Di Baldassarre et al., 2013a, 2014). The second approach also starts with a narrative understanding of the situation, but uses this to distinguish patterns for deriving causal relationships (Di Baldassarre et al., 2013b, 2015; Barendrecht et al., 2017). They thus scale down the complex realities to elaborate trends and casual relationships that are captured in mathematical models (Elshafei et al., 2014). A summary of the main contributions to and advances in socio-hydrology of the recent years is presented in chapter 2 of this thesis.

As noted, the hope is that a better understanding of the dynamics between hydrological and social processes in floodplains can improve policy making aimed at reducing risk and vulnerability and help selecting the best adaptation or mitigation options. Di Baldassarre et al., (2013a, b) identified two broad patterns of society-river interactions; (a) adapting to the river and (b) fighting the river (Figure 1.2). In this broad distinction, floodplains are expected to either demonstrate one or the other pattern at one point in time, while allowing shifts over time from adaptation to levee effect (Di Baldassarre et al., 2013b). Hence, the classification categorises a floodplain as having one dominant socio-hydrological pattern ('adapt' or 'fight').

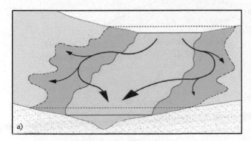 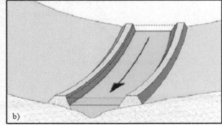

Figure 1.2: Schematic of human adjustments to flooding – (a) adaptation: settling away from the river, and (b) fighting: raising levees or dikes (Di Baldassarre et al., 2013b).

To enhance the knowledge on human-water interactions, this PhD research engages with this categorisation between 'adapt' or 'fight' and sets out to test, refine and further develop it for the specific case studied, using the actual data for specific cases. To do this, the study explores the hydrological and social processes along the Jamuna River and its floodplain and looks at the dynamic interactions between them. The main hydrological and social processes for Bangladesh have been identified through a literature review, outcomes and details of which are presented in the following chapters 2-5. Flooding and riverbank erosion have been identified as the most significant hydrological processes for the case, as these strongly affect the social processes along the Jamuna floodplain of Bangladesh (Hossain, 1984; Rasid and Paul 1987; Haque and Zaman, 1989; Hutton and Haque, 2004; Best, Ashworth, Sarker, & Roden, 2007; CEGIS, 2014). The study documents and traces how these hydrological processes generate and interact with demographic shifts, resettlements, human migration, urbanization processes, changes in governance, land use changes, changes in community resilience, changes in coping strategies, livelihood changes, human interventions were examined amongst the social processes (Elahi, 1972; Islam, 1976; Khan, 1982; Hutton and Haque, 2004; Hofer & Messerli, 2006; Best, Ashworth, Sarker & Roden, 2007; RMMRU, 2007; Penning-Rowsell et. al., 2012; Black et al., 2013).

An initial conceptualisation for this study links the dynamics of current hydrological and social processes for the Jamuna river floodplain (Figure 1.3). This model is based on previous research in Bangladesh (Ferdous, 2014) and on the hypotheses developed by Di Baldassarre et al. (2013b) for floodplains in general.

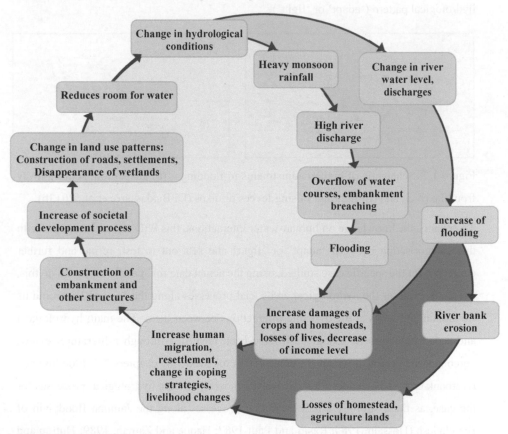

Figure 1.3: Initial conceptual model for socio-hydrological dynamics of Jamuna floodplain of Bangladesh.

According to this initial conceptualisation, an increase of heavy monsoon rainfall determines the increase in the discharge of the Jamuna River. As a result, water floods the countryside thereby sometimes breaching embankments which in turn results in flood and riverbank erosion in that area. Flood and riverbank erosion are provoking damage to crops and homesteads, the loss of lives, decreases in income level which may lead to human migration, resettlement, a change in coping strategies, livelihood changes etc. People may also be constructing or upgrading embankments and other structures to prevent future damage from flooding and river bank erosion. Floods are co-shaping the rate and nature of societal development by inducing changes in land use patterns, including the construction of roads, settlements etc. Yet, as a result of such changes in land use pattern, the room for water may also reduce, provoking further changes in hydrological conditions. Capturing the dynamics of these endless spirals of mutual influence between hydrological and social processes over time is the purpose of socio-hydrology in general. This study wants to do this for a specific part of the Jamuna floodplain.

1.4 RESEARCH OBJECTIVES

The main objective of this research is to better understand the interactions between physical processes and societal processes along the Jamuna River in Bangladesh as temporally dynamic and spatially diverse combinations of fighting with water and living with water.

The specific objectives of the research are as follows:

(1) To capture the different socio-hydrological patterns that result from different societal choices in different physical conditions on how to deal with rivers, floods and erosion in the Jamuna River floodplain in Bangladesh.

(2) To explore to what extent the levee effect can be distinguished in the Jamuna River floodplain in Bangladesh.

(3) To explore what outcomes result from adjustments to flooding and erosion in the Jamuna River floodplain in Bangladesh.

(4) To explore the interplay of flood vulnerability and structural protection levels in the Jamuna River floodplain in Bangladesh.

1.5 THESIS OUTLINE

The research objectives are achieved by means of the combined spatial and temporal analysis of social and physical data on a short stretch of the Jamuna floodplain near Gaibanda and Jamalpur of Bangladesh. Primary data were collected through 900 household surveys and 15 focus group discussions. Secondary data (river water level, erosion, population, satellite images) were collected from governmental agencies. The dynamics of hydrological processes and social processes have been analysed by statistical analyses and mapping with the aid of GIS.

In Chapter 2 the concept 'socio-hydrological spaces' is proposed. This concept captures the different socio-hydrological patterns that result from different societal choices on how to deal with rivers, floods and erosion. Chapter 2 also presents evidence for the existence of such spaces in the case study area. The concept of socio-hydrological spaces can be used both in generic models as well as in specific case studies to help understand the dynamics and co-evolution of coupled human-water systems. We identified three distinct socio-hydrological spaces in our study area along the Jamuna River; (i) areas protected with flood embankment; (ii) floodplain outside the embankment and chars (river islands) and (iii) areas with natural levee. We show that the socio-hydrological dynamics of these spaces look different depending on the government's flood risk management practices, since they shape behaviour, including social & individual adjustments to flooding. Chapter 2 addresses Objective 1 of the research.

Chapter 3 explores whether a levee effect can be distinguished between areas with different protection levels that are otherwise homogeneous in the case study area. The floodplain along the Jamuna River presents different flood protection measures on the two river banks. There is a man-made levee parallel to the right bank (the Brahmaputra Right Embankment, BRE), but no similar structural investment was made on the left bank, leaving it unprotected except by a natural levee. Furthermore, at the time of study the BRE showed different characteristics along its length, with stronger (reinforced with concrete and maintained) levees in the southern part, and weaker (unmaintained) levees in the northern part. Levees protect floodplain areas from frequent flooding, but they can also paradoxically contribute to more severe flood losses. This is known in the literature as the 'levee effect', which was first discussed by Gilbert White in the 1940s. We compare two data sets of two areas to see if we could distinguish a levee effect. The first set comprises two rural areas (protected rural area vs unprotected rural area), the second set comprises two urban areas (Gaibandha town with weak levee vs Sirajganj town with strong levee). By comparing rural areas with and without levees and urban areas with a strong and a weak levee, we have shown how levees contribute to relatively more economic growth, investment, agricultural incomes, etc. and lower year-by-year damages (but higher damages with extreme floods), etc. Chapter 3 addresses Objective 2 of this research.

In Chapter 4, we describe how people are living with flooding along the Jamuna floodplain in Bangladesh and explore how this affects their livelihoods. In this study, the distinction between coping and adapting is crucial. Coping includes long-term adjustments where the nett result can be a decline in socio-economic conditions, while adapting means that people continue to live as well, or better, than before, and are as well or better able to cope with future events. In this interpretation, adjustment to environmental conditions can result in adaptation but this is not necessarily the case. Both coping and adapting require the mobilisation of a variety of technologies and social or institutional changes, which we define as adjustments. The actions and strategies for adjustment may be the same, but have a different result: coping or adapting. In this chapter, we show how adjustments to environmental conditions can have negative outcomes, for example leading to impoverishment. We explore how these outcomes change depending on the socio-hydrological spaces presented in chapter 2, and also depending on economic status. To study the latter, we developed a classification that includes both current wealth and income. Chapter 4 addresses Objective 3 of this research.

The focus of Chapter 5 is the interplay of flood vulnerability and structural protection levels in Bangladesh. The construction or reinforcement of levees can attract more assets and people in flood-prone area, thereby increasing the potential flood damage when levees eventually fail. Moreover, structural protection measures can generate a sense of complacency, which can reduce preparedness, thereby increasing flood mortality rates. We explore these phenomena in the Jamuna River floodplain in Bangladesh. Our results support consolidated theories about the interplay between levels of structural flood protection, people and assets exposed to flooding, and social vulnerability to flooding. Chapter 5 addresses Objective 4 of this research.

A synthesis of the research results is presented in Chapter 6, as well as a discussion of possible policy implications of the findings.

2

SOCIO-HYDROLOGICAL SPACES IN THE JAMUNA RIVER FLOODPLAIN IN BANGLADESH

This chapter is published as:

Ferdous, M. R.; Wesselink, A.; Brandimarte, L.; Slager, K.; Zwarteveen, M. and Di Baldassarre, G. (2018). Socio-hydrological spaces in the Jamuna River floodplain in Bangladesh. Hydrol. Earth Syst. Sci., 22, 5159-5173, https://doi.org/10.5194/hess-22-5159-2018.

Abstract

Socio-hydrology aims to understand the dynamics and co-evolution of coupled human–water systems, with research consisting of generic models as well as specific case studies. In this chapter, we propose a concept to help bridge the gap between these two types of socio-hydrological studies: socio-hydrological spaces (SHSs). A socio-hydrological space is a geographical area in a landscape. Its particular combination of hydrological and social features gives rise to the emergence of distinct interactions and dynamics (patterns) between society and water. Socio-hydrological research on human–flood interactions has found two generic responses, "fight" or "adapt". Distilling the patterns resulting from these responses in case studies provides a promising way to relate contextual specificities to the generic patterns described by conceptual models. Through the use of SHSs, different cases can be compared globally without aspiring to capturing them in a formal model. We illustrate the use of SHS for the Jamuna floodplain, Bangladesh. We use narratives and experiences of local experts and inhabitants to empirically describe and delimit SHS. We corroborated the resulting classification through the statistical analysis of primary data collected for the purpose (household surveys and focus group discussions) and secondary data (statistics, maps etc.). Our example of the use of SHSs shows that the concept draws attention to how historical patterns in the co-evolution of social behaviour, natural processes and technological interventions give rise to different landscapes, different styles of living and different ways of organising livelihoods. This provides a texture to the more generic patterns generated by socio-hydrological models, promising to make the resulting analysis more directly useful for decision makers. We propose that the usefulness of this concept in other floodplains, and for other socio-hydrological systems than floodplains, should be explored.

2.1 INTRODUCTION

The hydrological sciences community has recently launched socio-hydrology as one of the research themes of the current scientific decade of the International Association of Hydrological Sciences (IAHS) "Panta Rhei – Everything Flows" (2013–2022), which aspires to "advance the science of hydrology for the benefit of society" (Montanari et al., 2013, p. 1257). Socio-hydrology aims to understand the dynamics and co-evolution of coupled human–water systems (Sivapalan et al., 2012). In traditional hydrology, humans are either conceptualised as an external force to the system under study, or taken into account as boundary conditions (Milly et al., 2008; Peel and Blöschl, 2011). In socio-hydrology, human factors are considered an integral part of the system. Understanding such coupled system dynamics is expected to be of high interest to governments who are dealing with strategic and long-term water management and governance decisions (Sivapalan et al., 2012).

As in any newly defined research area, socio-hydrology researchers are looking to determine how to implement their shared goal. This has resulted in a number of overview or position papers (e.g. Blair and Buytaert, 2016; Sivapalan, 2015; Pande and Sivapalan, 2017; Sivapalan and Blöschl, 2015; Troy et al., 2015) as well as several case studies (e.g. Gober and Wheater, 2014; Kandasamy et al., 2014; Liu et al., 2014; Mehta et al., 2014; Srinivasan, 2015; Mostert, 2018). In the discussions, the use of conceptual and deterministic models to analyse concrete situations is an important topic. As in any attempt to produce insights that transcend specific cases, methods of abstraction from reality to find causal relationships and stylised equations (generalisation) are sometimes difficult to reconcile with more detailed representations of what is happening in a specific location (Blair and Buytaert, 2016). While enabling global comparison by using data sets from different locations, generic models unavoidably foreground some elements or dimensions of flood–society dynamics to the neglect of others (Magliocca et al., 2018). However, attempts to generalize from case-specific detailed models need to be looked at critically in terms of the comparability and commensurability of the modelled phenomena with what happens elsewhere: detailed causal relationships in one case do not usually correspond to those in other cases (e.g. Elshafei et al., 2014).

In this chapter, we focus on another, less formal way to capture socio-hydrological dynamics than causal relationships and models: patterns. Pattern detection is no new activity in socio-hydrology, because patterns are at the basis of the stylised representations (equations) in generic models. Historical patterns are foundational for a full understanding of generic as well as place-based models, and pattern finding reinforces the feedback between empirical studies and modelling studies. However, we propose a new socio-hydrological concept to operationalise the search for patterns in the messy reality of specific cases: socio-hydrological spaces (SHSs). Eventually, patterns found in cases may be formalised into causal relationships, but this does not necessarily have to be the goal. We contend that patterns in and by themselves are valuable research results, especially in policy development and where data are scarce (Section 2.8). The SHS concept will be defined and its implementation explained in Section 2.3. To illustrate how the concept can be used, we analyse human–flood interactions in the Jamuna floodplain, Bangladesh, making use of the two generic responses to flood risk "fight" or "adapt" that were found in earlier research on human–flood interactions (Section 2.2). In the Jamuna floodplain the differences between land and water are temporary and shifting, as is the size of the human population. The application of SHS allows the capture of the different socio-hydrological patterns that result from different societal choices on how to deal with rivers, floods and erosion, which in turn produce different living conditions and watery environments (Sections 2.4 and 2.5).

The detection of patterns in socio-hydrological relationships can be based on the interpretation of a combination of qualitative and quantitative data; it is therefore more feasible where quantitative data are scarce. This mid-level theorizing on the basis of empirically observable patterns was identified by Castree et al. (2014) as a desirable way forward in environmental research, as it makes it easier to link and translate model-deduced patterns with experienced realities. By providing locally relevant details and texture to more generically deduced patterns, SHS provides a useful methodological addition to the socio-hydrological understanding of floodplains. Its usefulness to other contexts such as irrigated catchments or urban water systems could also be investigated.

2.2 PATTERNS IN THE SOCIO-HYDROLOGY OF FLOODPLAINS: FIGHT OR ADAPT

One type of situation that is relatively well studied by sociohydrologists is the co-evolution of human societies and water in floodplains. After all, the existence of interdependencies between societies and their natural environment is particularly obvious in floodplains. Since the beginning of human civilisation, many societies have developed in floodplains along major rivers (Vis et al., 2003). In spite of periodical inundations, a distinct preference for floodplain areas as places to settle and live in stems from their favourable conditions for agricultural production and transportation, enabling trade and economic growth (Di Baldassarre et al., 2010). Yet, floodplain societies have to learn how to deal and live with periodic floods and the relocation of river channels by erosion and deposition (Sarker et al., 2003). In general terms, floodplain societies do this by evaluating the costs of flooding and erosion against the benefits that rivers bring, and deciding whether to try to mitigate the risks by defending themselves against floods ("fight"), or to live with floods ("adapt"), or any combination of the two (Di Baldassarre et al., 2013a, b). Whether and how societies can fight or adapt to flooding depends on the society's economic and technological possibilities. Therefore, "fight" and "adapt" are the two generic responses in the socio-hydrological dynamics of human–flood interactions. These combine differently in different contexts and locations, resulting in different sociohydrological patterns.

For flood mitigation, societies have usually relied on engineering measures like embankments or levees to prevent flooding, and bank protection and spurs or guide bunds stretching into the river to prevent erosion. These measures can be seasonal (temporary) or permanent and have a greater or lesser effect on flood prevention (Sultana et al., 2008). The construction of flood control measures might in turn alter the frequency and severity of floods, leading to a dynamic interaction between the river and the society living along side it (Hofer and Messerli, 2006). An alternative response to flooding is adaptation. In order to adapt to flood risks, societies may limit costly investments in property or make them movable, adjust cropping patterns or choose crops that can cope with flooding, or move away altogether if alternative locations for settlement are available. Even when flood protection measures are in place, residual risks may necessitate adaptation measures. This means that in any real situation the two responses of "fight" and "adapt" are usually found together in a site-specific configuration, depending on socioeconomic, institutional and natural conditions. We label the areas where the proportions are analogous due to similar conditions (SHSs; see Section 2.3).

The study of floodplains using a socio-hydrological approach has advanced rapidly in the last few years (Di Baldassarre et al., 2013a, b, 2015; O'Connell and O'Donnell, 2014; Viglione et al., 2014; Chen et al., 2016; Grames et al., 2016; Ciullo et al., 2017; Barendrecht et al., 2017; Yu et al., 2017). In this research, the two responses to flooding "fight" and "adapt" take centre stage. The overall aim of this work is to further understanding on "how different socio-technical approaches in floodplains are formed, adapted and reformed through social, political, technical and economic processes; how they require and/or entail a reordering of social relations leading to shifts in governance and creating new institutions, organisations and knowledge; and how these societal shifts then impact floodplain hydrology and flooding patterns" (Di Baldassarre et al., 2014, p. 137).

Two different methodologies for the study of floodplains can be broadly distinguished, in parallel with general trends in socio-hydrology found by Wesselink et al. (2017). The first approach presents a narrative representation of the floodplain's socio-hydrological system. The narrative is generally based on qualitative research, often informed by experiences and knowledge of local experts and inhabitants about histories of living with floods, but may also include statistical data, e.g. on trends. The resulting studies describe historical patterns in the co-evolution of river dynamics, settlement patterns and technological choices (Di Baldassarre et al., 2013a, 2014). Not all researchers who engage in this kind of study identify their work as belonging to socio-hydrology (e.g. Van Staveren and Tatenhove, 2016; Van Staveren et al., 2017a, b). In this qualitative research, the actual societal choices between "fight" and "adapt" are descriptively represented, without formalisation.

The second approach to studying socio-hydrological dynamics of floodplains focusses on the development and use of a generic conceptual model of human–nature interactions, which is subsequently expressed in terms of differential equations (e.g. Di Baldassarre et al., 2013b, 2015; reviewed in Barendrecht et al., 2017). This second approach also starts with a narrative understanding of the situation, in which patterns are key for deriving causal relationships. These narratives narrow down complex realities to a selection of phenomena and elaborate trends and causal relationships that are subsequently captured in mathematical models (see Elshafei et al., 2014, for a clear example of the role of narratives). Generic models aim to explain the feedback mechanisms that produce certain phenomena (often paradoxes or unintended consequences) that have been observed in many places around the world. For example, the stylised models of human–flood interactions introduced by Di Baldassarre et al. (2013b) use a mathematical formalisation of a fundamental hypothesis: the levee effect (White, 1945) is explained by a decrease in risk awareness when flooding becomes less frequent because of the introduction (or reinforcement) of structural protection measures. This generic model has been used to explore and compare alternative scenarios of floodplain development (Di Baldassarre et al., 2015; Viglione et al., 2014). Current research includes further refinement (Grames et al., 2016; Yu et al., 2017) or comparison of this generic model to actual data for specific cases (Ciullo et al., 2017; Di Baldassarre et al., 2017). Yet, as societal responses to hydrological changes (including flood occurrences) are "very complex and highly unpredictable as it strongly depends on economic interests and cultural values" (Di Baldassarre et al., 2015, p. 4780), formalisation is challenging.

2.3 SOCIO-HYDROLOGICAL SPACES DEFINED

To reflect pattern detection as intermediary activity between modelling and reality, we define a socio-hydrological space in two ways: from the empirical observations, which may include quantitative data but also general contextual knowledge ("bottom-up"), and from the conceptual models of the general patterns found in human–flood interactions ("top-down"). Starting from the empirical observations captured in quantitative and qualitative data, we define a sociohydrological space as a geographical area in the landscape with distinct hydrological and social features that give rise to the emergence of distinct interactions and dynamics between society and water. Starting from the generic patterns captured in conceptual modelling, a socio-hydrological space is the empirical expression of a specific combination of generic responses (here: fighting and adaptation dynamics) in a geographical area that is distinct from the neighbouring one. Importantly, both definitions apply simultaneously and are operationalised in an iterative manner to study the sociohydrology of an area as shown in the example for the Jamuna flood plain (Sections 2.4 to 2.6).

Using SHS in the analysis of socio-hydrological dynamics helps to make the necessary intermediary step between the messy and many details used to characterise a specific location (space) and the stylised abstraction of generic models. With the proposal of SHS we are looking for a middle ground where we preserve the variability of reality and the unpredictability of human behaviour and decisions, not force-fitting these into a model, while at the same time recognising patterns (due to combinations of similar or comparable fight and/or adapt responses). We thus propose that SHS can serve the function of a lens through which to view and filter the complex reality of specific cases, in order to find patterns in human–water interactions. Such patterns can then be compared and contrasted to patterns in other locations to see if further generalisation towards generic models is possible. The use of SHS invites the researcher to have an open mind to the existence of expected or unexpected patterns in the location under investigation, using a thorough understanding of the specifics of this location in terms of society, history, economics, natural systems, technical interventions etc. Insights from one location can then be compared to analyses of other cases in order to explore whether the same or different patterns occur, and for the same reasons. These patterns can then be generalised through a more formal conceptualisation of socio-hydrological systems, whereby the existing conceptual models may be taken as a starting point. On the one hand SHS thereby relates to a specific space; on the other hand it helps to find general patterns of human–water interactions, which means that use of SHS to analyse different cases enables global comparison.

It is interesting to note that some of the earlier sociohydrological research on floodplains can be said to implicitly employ something resembling the SHS concept (Figure 2.1). In their study, which is partly based on the Po floodplain, Di Baldassarre et al. (2013a, 2014) identify two patterns of society–river interactions. In the "adaptation effect" pattern the use of flood defence technology is limited, resulting in frequent flooding that is in turn associated with decreasing vulnerability (see also Kreibich et al., 2017). The "levee effect" pattern results when flood protection structures lead to less frequent but more severe flooding, which is in turn associated with increasing vulnerability (Di Baldassarre et al., 2015) (already identified by White, 1945; see also Kates et al., 2006). These two patterns can be rendered in terms of SHS, yielding the following classification:

a) the "adaptation space", where frequent flooding results in less economic development and lower population density and other human adjustments;

b) the "fighting space", where flood protection structures lead to less frequent but more severe flooding, more economic development and higher population density and other human adjustments.

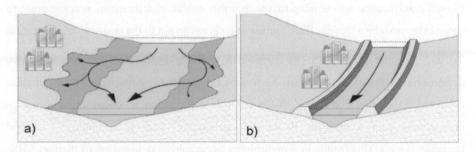

Figure 2.1: Schematic of human adjustments to flooding – (a) adaptation: settling away from the river, and (b) fighting: raising levees or dikes (after Di Baldassarre et al., 2013b).

In these first conceptualisations, one floodplain is assumed to show one or the other pattern at one point in time, while allowing shifts over time from adaptation to levee effect (Di Baldassarre et al., 2013a). This classification categorises a floodplain as having one single socio-hydrological pattern ("fight" or "adapt"). Di Baldassarre et al. (2015) then classify several floodplains worldwide in one of the two patterns. For example, they classify Bangladesh as a whole into the "adapt" type. However, it turns out that several sections of the floodplain in Bangladesh are protected by an embankment (see Section 2.5), with residual flood risks giving rise to adaptation behaviour. Similarly, their classification of the Rhine floodplain in the Netherlands as "fighting floods" holds in general, though in several places adaptation is being experimented with (Wesselink et al.,

2007; Van Staveren and Van Tatenhove, 2016). In the same country, the Meuse valley was classified as "adaptation" although embankments have been added to protect built-up areas (Reuber et al., 2005; Wesselink et al., 2013). As the goal of generic models is to describe decadal dynamics at large scale (Di Baldassarre et al., 2013b), they can only capture the main phenomena in large areas, such as a whole floodplain (in time or space) or a river basin. Instead, SHSs induce the researcher to further refine the analysis of human–flood interactions from the generic to the more local where, for example, both responses may coexist at one time in specific proportions. In this way, SHSs allow more specific and detailed representation of the reality of these interactions, while still enabling comparison between cases by referring to generic patterns. In what follows, we illustrate how the concept can be used in a more detailed and refined analysis of the Jamuna floodplain in Bangladesh. We show how its use can provide nuances to the broad-sweep overall classification by showing that within this overall characterisation some areas to some extent exhibit a "levee effect", while other areas do not fit the two-way classification.

To use the concept of SHS, we propose a two-step approach. First, the top-down definition of SHS guides the researcher to look for the generic patterns in the information collected about the study area. As noted this information is based on a thorough understanding of a specific floodplain (geography, history, technology, societal occupation etc.). This results in a preliminary geographical delineation of distinct SHSs and their qualitative descriptions by means of narratives, schematised drawings, maps etc.; these results have the function of being hypotheses in the next step. In the second step, quantitative data analysis is employed to confirm, reject or correct these initial hypotheses; that is, this analysis provides the data-driven (bottom up) delineation of SHS. If the classification is not statistically significant, merging or splitting of categories should be considered as well as redrawing the boundaries (repeat step 1). However, this adjustment should always be based on arguments based on a good understanding of the floodplain, since statistical significance by itself does not explain socio-hydrological dynamics.

Similar research methods were used before in sociohydrology, e.g. geo-statistics to study the interaction between river bank erosion and land use (Hazarika et al., 2015), or so-called data-driven narratives (Treuer et al., 2017) and the pairing of statistical analysis and narratives (Hornberger et al., 2015; Mostert, 2018). While the combination of narrative and statistical methods that we use is therefore not new, their application to SHS enables single case studies to be transcended in the search for more generalisable patterns.We could therefore envisage that the methods used in Step 2 (see Section 2.4.3) could be different, as long as they contribute to the goal of identifying and validating SHS.

The following case study demonstrates how the SHS approach can be used. Our goal is not to include all available data to provide an exhaustive analysis, but to show how SHS help to detect and understand socio-hydrological dynamics. The socio-hydrological characteristics and data availability guide the choice of methods in our socio-hydrological analysis of a part of the Jamuna floodplain in Bangladesh. In other circumstances the application of SHS will likely entail different variables and methods.

2.4 RESEARCH APPROACH

2.4.1 Case study area

The delta where the Ganges, Brahmaputra and Meghna rivers meet the sea in the Bay of Bengal encompasses 230 river channels and covers most of Bangladesh (Mirza et al., 2003). It is the largest delta in the world, draining almost all of the Himalayas, the most sediment-producing mountains in the world (Goodbred et al., 2003). The flows of the three rivers add up to an average of 1 trillion cubic metres of water per year and 1 billion tonnes of sediment per year. The sediment load is very high, resulting in very dynamic river channels (Allison, 1998). In the early 18th century, the main course of the current Jamuna was flowing through what is now the Old Brahmaputra, to the east of the Jamuna. Sometime between 1776 and 1830 the course of the Brahmaputra shifted from east to west, and the "new" river was given the name Jamuna. Since then, the Jamuna has shown progressive westward migration and widening, meanwhile transforming from a meandering river to a braided one (CEGIS, 2007). The Brahmaputra Right Embankment (BRE) was constructed on the west bank of the Jamuna in the 1960s to limit flooding and increase agricultural production, and also to try to stabilise the position of the river, the latter with limited success despite the addition of groynes and spurs.

25

Bangladesh is a very densely populated country with more than 140 million of people (964 persons per km^2). Around 80% of the population lives in floodplain areas (Tingsanchali and Karim, 2005) and depends on agriculture and fisheries (BBS, 2013). In the monsoon season, 25%–30% of the floodplain area is inundated every year (Brammer, 2004). These "normal" floods are valued by rural inhabitants because they are beneficial to the fertility of the land and provide ecosystem services (fish stock) and transportation possibilities (Huq, 2014). According to the classification by the Flood Forecasting and Warning Centre, which categorises flooding events as normal, moderate and severe based on flood duration, exposure, depth and damage, extreme flood events were observed in 1954, 1955, 1974, 1987, 1988, 1998, 2004 and 2007 (FFWC/BWDB, 2017); the flood events since then were not judged to be extreme in the whole country, but in north-western Bangladesh, which includes our study area, 2016 and 2017 were also extreme (FFWC/BWDB, 2017). Throughout the years, successive governments have implemented several flood control measures to protect agriculture and populations from floods (Sultana et al., 2008).

Riverbank erosion is associated with flooding in many areas of the country. The extremely poor people who live on the chars (islands in the big rivers) are most exposed to and affected by flood hazards and riverbank erosion. During the period 1973 to 2015, the net erosion was 90,413 ha and the net accretion 16,497 ha along the 220 km long Jamuna (CEGIS, 2016). Every year about 50,000 to 200,000 people are displaced by riverbank erosion, although they usually find another place to settle nearby in the area (Walsham, 2010). Hence, it is clear that hydrological processes (flooding and riverbank erosion) play a vital role in the way people in Bangladesh organize their lives, as manifested, among other things, in patterns of migration, livelihoods and land use.

To understand these relationships between river and people better, this study focusses on a small area along approx. 30 km of the Jamuna River in the north of Bangladesh (Figure 2.2). The total area is about 500 km^2 and the total population is approx. 0.36 million (BBS, 2013, BBS, 2014a). The case study area includes parts of Gaibandha district and parts of Jamalpur district (Figure 2.2). The total width of the case study area is around 24 km, of which the braided river bed takes approx. 12–16 km; this includes many inhabited river islands (chars) that flood with varying frequency (from every year to only with severe floods). The maintenance of the BRE in the study area has been sporadic. When constructed, the average height was 4.5 m, width 6m and slope 1 : 3 on both sides (CEGIS, 2007). Though extreme discharges could not overtop this embankment, breaches have occurred which caused catastrophic floods and damages (RBIP, 2015). In the 2016 flood (observed during the field survey), the BRE was breached in Gaibandha district, resulting in a large area being flooded. On the left bank there is no human-made protection, but there is a natural levee that has been deposited by the river.

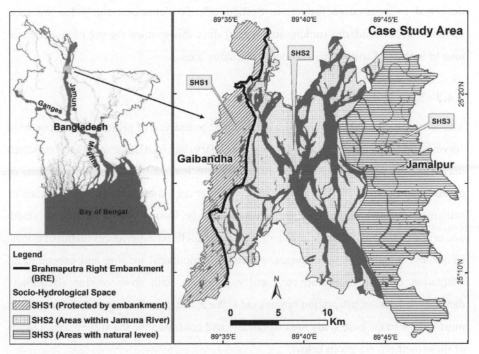

Figure 2.2: Bangladesh map with case study area and SHS.

2.4.2 Step 1: preliminary identification of SHS and classification of areas

Throughout the fieldwork period needed to collect the primary data described in Section 2.4.3 below, a detailed knowledge of physical, technical and social conditions of the area was accumulated by the author of this thesis. In collecting this information, he built upon and was guided by his personal knowledge as a resident in a nearby area, as well as by 10 years of professional experience throughout Bangladesh as a water engineer charged with flood forecasting and training residents on using flood and erosion forecasts. Since flood

27

control measures were only developed along some rivers (see Section 2.4.1 above), the study area is characterized by different degrees of protection. In addition to these human-made structures, different geomorphological conditions influence local flood frequency and extent as well as the extent of river bank erosion. Inhabitants adapt to these physical conditions, which is apparent, for example in private investment levels and cropping patterns, but also in public investment, e.g. in schools and roads. These qualitative observations formed the basis for distinguishing three SHSs in the landscape, which are described in a narrative fashion in Section 2.5. To demarcate the SHSs we used administrative boundaries (unions and mauza) since this enabled the use of Government data in Step 2; 15 unions are included in the study area.

2.4.3 Step 2: evidence

The demarcation of SHSs was validated through the analysis of primary data (household surveys and focus group discussions) and secondary data (statistics, maps etc.) collected during the dry seasons of 2015 and 2016. The principal set of primary data consists of approx. 900 questionnaires dealing with several themes: general information (location of settlement and agricultural land, main occupation, age, income and expenditures, wealth and origin of the households), information on different flood experiences (depth of floods, frequency, duration, flood damages, effects on agricultural income and expenditures, adaptation options, migration etc.) and experiences with river erosion (frequency, damages, migration, adaptation options etc.). We also set up focus group discussions in most unions in the case study area to validate and contextualise the survey data. Details of these methods are given below.

A cross-sectional method was used to gather the primary data of the case study area. Cross-sectional research involves using different groups of people, both male and female (farmer, fisherman, day labourer, service holder etc.) who differ in the variables of interest but share other characteristics, such as socio-economic status and ethnicity. We aimed to collect approximately the same number of surveys in each of the three SHSs. Due to the rural setting, most of the respondents were farmers. We introduced an age bias because we wanted to collect historical information on flooding, riverbank erosion, livelihood etc. The household surveys were implemented with a combination of purposive sampling and quota sampling. Purposive sampling is a method where individuals are selected because they meet specific criteria (e.g. farmer, fisherman, day labourer). The quota sampling method selects a specific number of respondents with particular qualities (like that a farmer's age should be 40 or above). We used the Raosoft sample size calculator to determine the required sample size for the surveys by union. This calculator allowed values to be entered, including an acceptable margin of error, response distribution, confidence level and size of the population that is to be surveyed. We accepted a 5% margin of error with 95% confidence level to determine the sample size, which is 1% of households (863 household surveys) in the case study area. The questionnaire for the survey is provided in the Appendix A.

In addition, we performed 12 focus group discussions in the case study area – 4 meetings in different unions in each SHS. About 20 participants were present in each of the meetings. Participants were selected based on occupation and location of the households, guaranteeing a uniform spread over the union area. The topics of the discussions were as follows: how flooding is affecting livelihoods; what household coping strategies are used in relation to flooding, for example changing occupation or raising homesteads; migration patterns; community interventions against flooding; river bank erosion and household coping strategies; community interventions against riverbank erosion; governmental initiatives against flooding and riverbank erosion etc. The agenda of the focus group discussions is provided in the Appendix B.

We also collected secondary data like time series satellite images to analyse the morphological dynamics of the Jamuna and census population data to analyse population density from different governmental and non-governmental organisations of Bangladesh. Results of Step 2 are discussed in Section 2.6.

2.5 RESULTS STEP 1: IDENTIFICATION OF SOCIO-HYDROLOGICAL SPACES ALONG THE JAMUNA RIVER

As noted, in our study area along the Jamuna, three distinct socio-hydrological spaces were identified. SHS1 covers the areas protected by the BRE (on the west bank), SHS2 covers the char areas (in the river bed) and SHS3 includes areas with a natural levee (on the east bank). These are depicted in a schematised fashion in Figure 2.3 and described by means of narratives below.

2.5.1 Areas protected with flood embankment (west bank) (SHS1)

This socio-hydrological space is protected from regular annual flooding, the so-called "normal floods", by the embankment along the main river Jamuna (BRE) and along some smaller Jamuna tributaries. However, different parts of the area are still frequently inundated with excess rainwater, due to their low elevation and limited drainage capacity. Further, a few small rivers (Ghagot and Alai) inundate unprotected areas yearly in the western part of the area. Because the BRE effectively protects the area against all but the largest riverine flooding from the Jamuna, inhabitants feel confident enough to invest in businesses and homesteads. In the study area, Gaibandha district, the BRE is not very well maintained, so the BRE sometimes breaches. Inhabitants build their houses on artificially raised platforms – often several metres above ground level – to reduce their vulnerability to the resulting floods. River bank erosion in this area is not widespread, but does occur in several locations. SHS1 therefore shows a combination of the "fight" and "adapt" patterns.

2.5.2 Floodplain outside the embankment (west bank) and chars (SHS2)

This is a very dynamic environment. The Jamuna is a braided river, where multiple channels criss-cross within the outer boundary of the river. When considered over decades, the outer boundary is moving in a westward direction (CEGIS, 2007). The "chars" – or river islands – are also moving, progressing or disappearing, due to local erosion processes. Chars have different ages, which have a direct relation to the height level. As the river still deposits sediment on chars, some older chars have higher elevations than the areas in SHS1 and have been shown to remain dry in extreme flood conditions. If a newly developed char does not erode immediately, it is first colonised by grass, which accelerates deposition of silt during the next flooding. Subsequently, people start to occupy the char, planting fast-growing trees and laying out agricultural fields. In the course of time, all kinds of facilities like schools, mosques, small shops, bazaars etc. are established. Since the chars are not stable, most of the houses built in the chars are semipermanent and easy to take apart and move. House types are kutcha (wood, straw and bamboo mats) or jhupri (straw). Many people raise the plinth levels of their houses to avoid flood damages, but this is not very effective.

On the chars inhabitants regularly face damages from flooding and river bank erosion to agricultural land and crops and their homestead, often leading to complete destruction. Temporary migration during the flood season to safer places, for example the embankment or on railway lines, is therefore very common. Permanent resettlement occurs only when the land that people live and farm on simply disappears, although they usually find another place to settle on a nearby char when floodwaters have receded. People also sometimes change their occupation temporarily or permanently. As char dwellers' lifestyles are defined by flood and erosion, they appear to be able to cope with the harsh conditions. Yet, most of them become poorer through time, because of landlessness, unreliable and changing sources of employment, and frequent temporary migration or resettlement. SHS2 therefore shows only the "adapt" pattern.

2.5.3 Eastbank (areas with natural levee) (SHS3)

The natural levee on the east bank of the Jamuna protects this area from about half of the annual riverine flooding; flooding occurs more frequently than in SHS1. A few areas are flooded by smaller rivers like the Old Brahmaputra and Jinjira. High water levels in these rivers sometimes occur independently of high water levels of the Jamuna, as these are not part of the same drainage basin. River bank erosion is conspicuous in this area. Even though the river as a whole shows a gradual westward shift, due to the presence of highly erodible bank materials on the left bank erosion is still severe in SHS3. For example, 75 ha of land eroded in 2015 in this area, of which 4 ha contained housing (CEGIS, 2016). Inhabitants take the initiative to build small spurs and bank protection, made from bamboo and wood, to try to stop erosion. However, while these encourage sedimentation at a local scale, they are not sufficient to stop largescale erosion. As in SHS1, most houses are built on artificially raised mounds, substantially reducing the potential for flood impacts. Flooding and riverbank erosion cause damage to agriculture, homesteads and businesses, in turn impoverishing people. As in SHS1 and SHS2, migration is one of the coping strategies, while households also adapt their cropping pattern to accommodate flooding and cultivate fast-growing crops after the flood season. SHS3 therefore shows a combination of the "fight" and "adapt" patterns, with more "adapt" and less "fight" than SHS1.

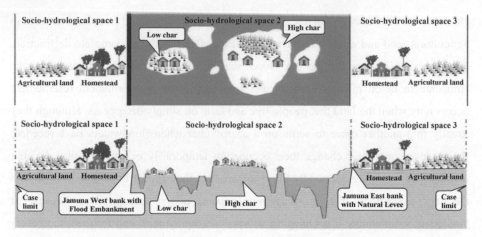

Figure 2.3: A typical planform and cross-section with distinct SHS along the Jamuna.

2.6 RESULTS STEP 2: EVIDENCE OF SOCIO-HYDROLOGICAL SPACES ALONG THE JAMUNA RIVER

Using the data described in Section 2.4.3, in this section we show that the three SHSs described above are significantly different. We only show the results for a limited number of variables: perceptions of the sources of flooding, flood frequency, flood damages, average household income and wealth, river bank erosion, migration and homestead types in the three identified SHSs. We performed statistical analysis Chi-square tests and ANOVA tests ($p < 0.05$) with these data for all analyses below (details are provided in Appendix C). In each case the data for the three socio-hydrological spaces were significantly different.

2.6.1 Perception of the sources of flooding

All respondents have experienced flooding in their lifetime, but their perceptions about the sources of flooding are different (Figure 2.4a). The main sources of flooding in SHS1 are excessive rainfall, neighbouring small rivers and the Jamuna (through breaching of the BRE), whereas the only source of flooding mentioned in SHS2 is the Jamuna, and for SHS3 they are the Jamuna, the Old Brahmaputra River and other smaller rivers. A significant number of people in SHS3 mentioned that the lack of embankment is one of the reasons for flooding, although they also mention excess discharges and river sedimentation.

2.6.2 Flood occurrence

When asked about their recollection of historical flood events (Figure 2.4b), in SHS2 people indicated that they experienced flooding every year. In both other spaces, this is roughly only once every 2 years. The unexpected relatively high flood frequency for the protected SHS1 may be attributed to the frequent failure of the embankment and to the fact that the area is flooded from the west by the Ghagot River, a tributary of the Jamuna.

2.6.3 Flood damage

The 1988 flooding was the most severe event for all three spaces, ranging from average damage of USD 800 per household in SHS3 to USD 1200 in SHS2 (Figure 2.4c). In other years, average flood losses were much lower. In 1987, damages in SHS1 were highest of the three spaces (USD ~ 700 per household). This may be attributed to poor drainage capacity in SHS1, as well as a lower average land elevation, resulting in deeper and longer water logging. Damages in 2007 and 2015 show little difference between the three SHSs (USD ~ 200 per household). It is interesting to observe that (apart from the 1988 event) flood damage in SHS2 is lower than damage in SHS1 and SHS3. This is not only because people there are generally poorer (Figure 2.4e), but also because people there are better adapted to frequent flooding (as they get flooded every year; see Figure 2.4b). Yet, while people in SHS2 have adapted to frequent flood events, this adaptation does not make them less vulnerable to big floods, such as the one of 1988 (see Figure 2.4c). This outcome was unexpected, and it would not be captured by any of the current models of human–flood interactions proposed so far.

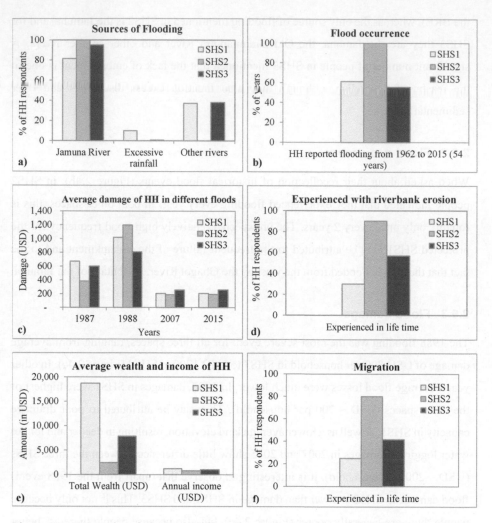

Figure 2.4: Comparison in between different socio-hydrological spaces (HH = household).

2.6.4 River bank erosion

Riverbank erosion is experienced in each SHS, but (as expected) mainly by inhabitants in the dynamic SHS2 (Figure 2.4d). However, erosion in SHS3 is also very high, with over 50% of the interviewed people having experienced it. In SHS1 expected rates are the lowest, but still considerable, as 30% have experienced it due to breaching of the BRE.

2.6.5 Average household income and wealth

The average wealth distribution (Figure 2.4e) clearly shows the economic differences between the households in the three SHSs. In the protected areas, people have much more wealth, on average about USD 19,000 per household, against approx. USD 2,500 in SHS2 and USD 8,000 in SHS3. Household wealth includes land (homestead, agricultural, other land), ponds, houses and housing materials, livestock, and portable wealth like savings, gold and silver. About 80% of the people in the case study area are farmers, so their income mostly depends on their agricultural production, complemented by remittances from migrant labour by family members for some families and from occasional day labour in agriculture, construction or fishing, or as rickshaw driver or van puller. Their starting position and subsequent losses depend to a large extent on where they live. The current situation is (much) worse for most households than in the past. As per our survey, in SHS1 large farms made up 7% of households (with land > 3 ha) in 1960 but after consecutive flooding events, this was reduced to only 2% in 2015 (Figure 2.5). Those who owned most land in the past (>3 ha) gradually saw a decline in their farm land to medium (1–2.99 ha) or small (0.2–0.99 ha), with some even becoming landless. There were only 16% landless households in 1960, but this increased to 28% in 2015.

In SHS2 and SHS3 a comparable pattern can be observed. The number of large farm households reduced from 18% to 1% and landless farmer households increased from 7% to 48% in SHS2. In SHS3 the proportion of large landowners reduced from 10% to 2% and that of landless farmers increased from 18% to 41%. More than 80% of the respondents from SHS2 reported that they could not recover from the losses due to flooding and riverbank erosion. Many of them have to change their occupation temporarily, and 3% of the respondents in SHS2 changed their occupation permanently from farmer to day labourer. This is less than the reduction in land ownership would suggest because landless farmers will try to rent land to be able to cultivate their own crops. If this is the case, they share crops with the land owners or pay a fixed amount per year.

There is a possibility that some respondents exaggerated reported losses in the hope that the research would help to mobilise funds. The focus group discussions clarified this issue.

They revealed that cropping patterns in SHS1, SHS2 and SHS3 are different. Respondents in SHS1 are cultivating three crops per year. In SHS3 people used to cultivate three crops in the past, but due to flooding, they now cultivate either two crops or only one crop per year, only in the dry season after floods have subsided. From the survey data it appears that in SHS1 only 15% of the respondents changed cropping patterns between the 1960s and the 2010s, compared to 53% in SHS3 and 40% in SHS2. A very small number of people have changed land use completely, for example from agriculture to homestead, from low elevation land to high elevation land by filling silts, or from agriculture to fallow etc.

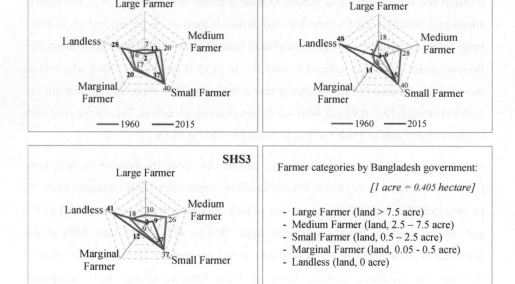

Figure 2.5: Agricultural land changes with time of the different types of farmer (% of HH respondents).

2.6.6 Migration

The population density census data for the three spaces show much higher densities in SHS1 than in SHS2 and SHS3. In SHS1 it is 1,500 people per km^2 (varying between 1,000 to 3,000 people per km^2 in the different villages in SHS1), while population density in SHS3 is 800 people per square kilometre (between 100 to 2000 people per km^2, the lowest figure being for very few villages adjacent to the east bank). It is lowest in SHS2 at 400 people per square kilometre (varying between 30 to 1,000 people per km^2) (BBS, 2013). The historical population data from 1961 to 2011 show that population density has increased in most of the unions, except in SHS2 (CPP, 1964; BBS, 1974, 1986, 1994, 2005 and 2013). Unfortunately, there are no official records of the exact number of people who migrate out of the area on a temporary or permanent basis. From our survey, we found that temporary or permanent migration is most frequent in SHS2, mostly to SHS1 and SHS3. From 1988 to 2015, 17% of respondents had migrated to SHS1 and 8% to SHS3.

The study shows that riverbank erosion (Figure 2.4d), more than flooding (Figure 2.4b), is one of the main drivers for relocation from a place of origin (Figure 2.4f). We found that 80% of the households in SHS2 had moved at least once. Most of them moved within 5 km, but in focus groups it was said that about 25% of people of that area had migrated away to other districts. About 68% of respondents were born in SHS1 and still live there, while 25% migrated to SHS1 from other places due to riverbank erosion. In SHS3, about 58% were born locally and the rest moved into the area, again mostly due to riverbank erosion. The respondents who relocated within the study area knew that their destination was flood-prone and at risk from riverbank erosion. However, the lack of available land is a major problem so they contend with sub-optimal conditions.

2.6.7 Homestead types

The construction types of houses vary among the spaces (Figure 2.6). Most of the pucca houses (well-constructed buildings using modern masonry materials) and semi-pucca or half-pucca houses (made of brick and tin) are within SHS1 and SHS3, where people feel comparatively safe against flooding and erosion. As a result, they invest more in their home. In SHS2 a high proportion of kutcha (wood, straw and bamboo mats) and jhupri (straw) houses is observed, since these are easy to take apart and move in case of flooding or erosion, and less costly to construct.

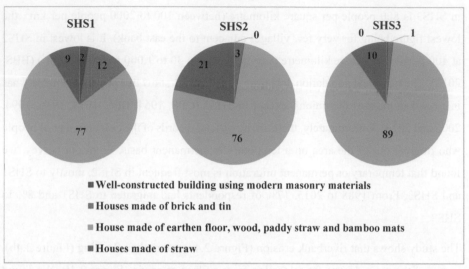

Figure 2.6: Homestead type of households.

2.7 DISCUSSION

Based on thorough in-depth knowledge of the natural, technical and social conditions of the study area in the floodplain of the Jamuna River in Bangladesh, we proposed distinguishing between three SHSs as the basic spatial units each with distinct socio-hydrological characteristics. Human–flood dynamics are different in each space, ranking from "adapt to floods" (SHS3), to more (SHS1) or less (SHS2) "fighting floods" in combination with "adapt to floods" to the extent necessary. We then proceeded to demonstrate, through statistical analysis of primary and secondary data, that the SHSs show significant differences in the following hydrological and social variables: perceptions of the sources of flooding, flood frequency, flood damages, average household income and wealth, river bank erosion, migration, and homestead types.

We thereby showed that there are good reasons to consider the three SHSs as distinct both from a narrative and from a statistical perspective, and that such a distinction provides a good starting point for further socio-hydrological analysis of human–flood dynamics of the area. Further research can reveal more details of the socio-hydrological feedback loops and resulting patterns in data within the three SHSs, for example by including urban areas. Our research is limited to rural inhabitants, and other patterns are likely to be revealed in SHS1 and SHS3 for urban areas, creating subdivisions within the spaces – or a reason to distinguish five instead of three spaces. The categorisation of the study area into spaces therefore depends on the focus of the study, but this does not invalidate the results. Rather, it shows that every abstraction, whether to find patterns or causal relationships, requires selective treatment of reality.

The issue of drawing boundaries around the SHS gives rise to another qualification. We started by outlining the boundaries of three SHSs based on the presence of distinct physical features in the landscape: the embankment on the west bank, the natural levee on the east bank and the riverbed in between. The exact boundaries were drawn on pragmatic grounds, using the administrative boundaries that best align with the physical features. These boundaries might show the approximate SHS in the present, but the boundaries of the physical and social systems are not fixed in time. The physical boundaries of the SHS are quite dynamic due to continuing bank erosion along both banks of the Jamuna (CEGIS, 2007). In particular, by analysing satellite images of the case study area from the late 1960s up to now, it appears that the west bank has been migrating westward and the east bank has been migrating eastward (Figure 2.7). As a result, the length-averaged width of the river has increased from 8.17 to 11.68 km (CEGIS, 2007). Since the construction of the BRE in the 1960s, many breaches have occurred due to river bank erosion, forcing relocation of the embankment in many places (RBIP, 2015). At the same time, due to erosion of the east bank the natural levee also moved somewhat over time. Thus, the physical boundaries between SHS1-SHS2 and SHS2-SHS3 are not fixed in time, while our statistical analyses assume that they are since they use the current physical boundaries. The social boundaries of the SHS in the Jamuna floodplain are also dynamic. Due to frequent relocations and migrations, the current inhabitants of the SHS may not have lived there throughout the study period since in every extreme event some migration occurs among the spaces (see Section 2.6.6). Therefore, the social boundaries between SHS1–SHS2 and SHS2–SHS3 are not fixed in time either, while our statistical analyses assume that they are since they relate to the current inhabitants. The dynamic nature of the boundaries of the SHS is unavoidable and indeed intrinsic to the highly dynamic socio-hydrology of the floodplain system. It is therefore important to remember that the SHSs are defined by their unique socio-hydrological characteristics compared with the surrounding area, not by their exact coordinates. For example, SHS2 is defined as a char within the river. If the river moves a kilometre and the char moves with it (or a different char forms), this does not change the definition of SHS2 as a char within the river. The same holds for the social boundaries, if one person moves to another SHS and adopts the strategies of that SHS, then the SHS does not change. Ideally, the data collection and analyses of time series in Step 2 would follow these shifting boundaries, but this will most likely not be possible due to data scarcity or time constraints.

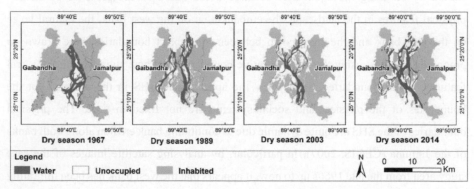

Figure 2.7: Time series dry season satellite images of the case study area.

2.8 CONCLUSIONS

We introduced the concept of socio-hydrological spaces (SHSs) and applied it to a floodplain area along the Jamuna River in Bangladesh. SHSs delineate areas where the interaction between social and hydrological processes show distinct characteristics, which in the case of floodplains can be classified as different combinations of two basic responses identified in the literature: "fight floods" or "adapt to floods". SHS are therefore primarily a research tool that helps to identify patterns in a specific case. However, when SHS are applied to other floodplains this will enable global comparison of human–flood interactions elsewhere. For example, similar SHSs to the ones found in the Jamuna floodplain are known to exist further down and upstream along the same river (known as Brahmaputra in India), so it would be worthwhile to compare socio-hydrological characteristics and analyse their differences and similarities.

Applying the SHS concept draws attention to the historical patterns in the co-evolution of social behaviour, natural processes and technological adoptions that give rise to different landscapes, different styles of living and different ways of organising livelihoods in specific geographical locations. The SHS concept suggests that the interactions between society and water are place-bound and specific because of differences in social processes, technological choices and opportunities, and hydrological dynamics. Such attention is useful anywhere in the world and also for other socio-hydrological systems than floodplains. It will be therefore be worthwhile to see whether SHSs can also be used to analyse physical processes other than floods, such as droughts, salt intrusion, irrigated catchments or urban systems.

The usefulness of SHSs does not only result from what it allows us to see, as explained above, but also from the relative ease of application in situations where data are too sparse to use fully deterministic models (which is the case nearly anywhere in the world). Compared with existing approaches in socio-hydrology, the concept allows an intermediary (narrative and/or statistical) position to be taken between complex realities and generic models. As such, it is argued that SHSs and generic models are complementary approaches with their respective advantages and disadvantages, making them useful for different purposes in different contexts.

Because SHSs are place-bound and can only be found (literally) on the ground, the use of SHSs forces the researcher to actually go to the field, talk to inhabitants and officials, and obtain a thorough understanding of the specifics of the location. This also means that the use of SHSs will make socio-hydrological analyses more policy-relevant. In terms of practical use, it can be added as an additional element to rapid rural appraisals, or other social assessments, to draw attention to how material conditions (hydrological and technical/ infrastructure) co-shape social situations. The application of SHSs is particularly useful to avoid broad-brush generalisations that do not take account of locality-bound problems due to the physical environment, without the need to interview every single household. SHSs are therefore useful for developing interventions under disaster management, but also other development goals. In summary, SHSs provide a new way of looking at and analysing socio-hydrological systems.

3

THE LEVEE EFFECT ALONG THE JAMUNA RIVER IN BANGLADESH

This chapter is published as:

Ferdous, M. R.; Wesselink, A.; Brandimarte, L.; Di Baldassarre, G. and Rahman, M. M. (2019). The levee effect along the Jamuna River in Bangladesh. Water International, https://doi.org/10.1080/02508060.2019.1619048.

Abstract

The levee effect refers to the paradox that the construction of a levee to protect from flooding might induce property owners to invest more in their property, increasing the potential damages should the levee breach. Thus, paradoxically, the levee might increase flood risk. The levee effect was observed for high-income countries. We analyze whether it can also be observed in a low income country, Bangladesh. In the Jamuna floodplain different levels of flood protection have existed alongside each other since the 1960s, so their effects can be compared.

3.1 INTRODUCTION

In 1945, Gilbert F. White wrote that 'floods are acts of God, but flood losses are largely acts of man. Human encroachment upon the floodplains of river accounts for the high annual toll of flood losses' (White, 1945, p. 2). Following this observation, White postulated the existence of a 'levee effect' in flood risk management and presented evidence for this effect in the United States. 'Levee' is a universal term for embankments (man-made or natural) that prevent floodwater flowing from a river to the surrounding areas; dikes are the man-made levees. In this chapter we will use 'levee', since White's hypothesis and subsequent research used this term. The levee effect is when the construction of levees to protect property from flooding induces property owners to invest more in their property, multiplying the risk should the levee breach or be overtopped. Of course, investment in property or economic capital may be exactly why the levee was constructed in the first place. With risk being defined as the product of the probability of events (here: flooding) occurring and the likelihood caused by damage of such events (Di Baldassarre, Castellarin, Montanari, & Brath, 2009; van Manen & Brinkhuis, 2005), the levee effect implies the paradoxical result that the construction of a levee can increase rather than reduce risk: while the frequency of flooding is reduced, the potential damage is magnified. Whether the risk indeed increases when a levee is constructed depends on the relative magnitude of the two factors, which will be different in different places and at different times (Tanoue, Hirabayashi, & Ikeuchi, 2016).

To reduce these risks, societies typically construct stronger levees, which again potentially increases the risk, and so on. The result is a lock-in of inescapable dependence on levees and ever-increasing expenditure for flood defence, with levees a condition for any inhabitation (Di Baldassarre et al., 2018; Logan, Guikema, & Bricker, 2018). This situation is perfectly illustrated by the Netherlands (Wesselink, 2007; Wesselink, Bijker, de Vriend, & Krol, 2007). The awareness of this lock-in has resulted in policy recommendations to 'go soft' in flood risk management, i.e., to focus on prevention, preparedness, emergency response and recovery after flooding, as for example in the EU Floods Directive (Gralepois et al., 2016; Hegger et al., 2016; Kreibich et al., 2017; Wesselink et al., 2015). Such 'soft' measures are believed to prioritize natural capital, community control, simplicity and appropriateness, while 'hard' engineering such as levees, groynes and revetments are capital-intensive, large, complex (and hence out of community control) and inflexible (Sovacool, 2011). This development to include more soft measures is a diversification that is needed in flood risk management; both hard and soft measures are needed for resilience to flooding. Therefore, trade-offs between the advantages and disadvantages of hard and soft measures should be considered carefully, as we will discuss in the concluding section.

Many studies have investigated the levee effect. Our literature survey yielded two types of study (see the following section). Most are single case studies to empirically test White's hypothesis in high-income countries, using historical records to show trends. A few conceptual studies use agent-based models or dynamic systems models where case study data illustrate simplified hypothetical situations. We found no research that compares adjacent areas at the same time, where social and physical characteristics are similar but flood protection levels differ substantially. Our unique case study, the floodplain along the Brahmaputra/Jamuna River (Bangladesh), enables just this. This floodplain presents different flood protection measures on the two river banks. The Brahmaputra Right Embankment (BRE), a man-made earthen levee, was built in the 1960s parallel to the right bank (BWDB, 1992), but no similar structural investment was made on the left bank, leaving it unprotected. Furthermore, at the present time the BRE shows different characteristics along its length, with stronger levees (reinforced with concrete, and maintained) in the southern part of the right bank, and weaker (unmaintained) levees in the northern part. These differences enable us to compare the effect of different protection levels within an otherwise homogeneous region. Also, our case offers a unique opportunity to test the levee effect in a low-income country.

After reviewing the past studies on levee effects, we present our study area and research approach, and then present and discuss the research findings. We conclude with a few thoughts on policy implications.

3.2 LITERATURE REVIEW: THE LEVEE EFFECT

White's (1945) publication has shaped the way flooding is perceived and revolutionized the methods by which risk and hazards are conceptualized more generally (Macdonald, Chester, Sangster, Todd, & Hooke, 2011). Indeed, White's thesis has inspired several studies in which the levee effect was investigated. Below, we briefly review this literature. To focus our review, we selected publications in Web of Knowledge where explicit reference is made to the levee effect in the title or topic, or White (1945) is cited. The themes covered are: investment or land use (i.e., economic development), risk perception, hydraulic conditions, potential damage or vulnerability, and finally risks. The term 'vulnerability' is often used in these studies, but not always defined explicitly. It often denotes 'expected damage', while in risk research vulnerability is more formally defined as the product of exposure and hazard (Mechler & Bouwer, 2015), where

- exposure is the economic capital that is potentially affected by hazardous events, and

- hazard is the likelihood and severity of events.

However, other authors define vulnerability as the 'likelihood of damage (including death or other undesirable consequences)' (Reilly, Guikema, Zhu, & Igusa, 2017), which includes both exposure and vulnerability as defined e.g., by Mechler and Bouwer (2015). We will adopt the definition, where exposure is included in vulnerability, since this is how it is interpreted in most of the literature we discuss below, if often implicitly. Understanding of vulnerability in the socio-ecological systems and development studies literature is different again, with vulnerability closely related to measures of resistance and resilience, besides exposure (Adger, 2006; Gallopin, 2006; Pelling, 2003).

In studies on the levee effect, economic development is most often studied. Burton, Kates, and White (1968) undertook research on human adjustments to natural hazards on the floodplains of the Mississippi River and the Rio Grande. They observed that farmers in protected areas were cultivating more valuable crops than those in unprotected areas. Shin, Hong, Kim, and Kim (2014) describe how inhabitants' behaviour changed after levee construction along three reaches of the Mankyung River, in South Korea. Levee construction led to a decrease of fallow land and an increase of urban and farmed land, which increases value and productivity. Eakin and Appendini (2008) also illustrate how the construction of levees in the Lerma Valley in Mexico rasied agricultural productivity and the economic value of farmland, but also increased vulnerability to flooding

Turning to urban and industrial development, Domeneghetti, Carisi, Castellarin, and Brath (2015) examined economic and population dynamics in the Po floodplain in Italy, finding that industrial investment expanded over the last century, particularly in floodplains protected by levees, but their analysis was not conclusive as to a causal relationship between the two. Collenteur, de Moel, Jongman, and Di Baldassarre (2015) came to a similar conclusion in their study of the impact of flood disasters on the spatial population dynamics in floodplain areas of the Mississippi River in the US. They found a positive correlation between the frequency of flood damage and population growth, but they could not show causality. In their historical study of flood risk management in New Orleans, Kates, Colten, Laska, and Leatherman (2006) showed that the construction of levees induced additional development in the city, but also produced the lock-in described above. Montz and Tobin (2008) similarly found that the construction of levees in Yuba County, California, promoted floodplain development, even though protection could not be guaranteed due to poor maintenance.

In the reviewed papers, the inhabitants' perception of flood risk (rather than the calculated flood risk) is usually seen as the intermediate step between the construction of levees and increased economic development and/or investment. Eakin and Appendini (2008) found that the presence of levees reduced farmers' perception of flood risk and led to increased investment. Ludy and Kondolf (2012) found that people in California believed that flood risk was completely eliminated by levees, though they are effective only against floods with a return period shorter than 100 years. This sense of safety fostered intense urbanization of flood-prone areas. Fox-Rogers, Devitt, O'Neill, Brereton, and Clinch (2016) studied how flood risk preparedness might be increased in the coastal and riverside town of Bray, in County Wicklow, Ireland. They found that the levee effect could occur even before levees are fully constructed, as the people living in the area felt that flood threat was being reduced. This in turn diminished flood preparedness. Burton and Cutter (2008) examined an additional social factor that increases potential damage: the composition of the population in areas at risk from flooding due to levee failures in the Sacramento–San Joaquin Delta in California. They found that as cities expand and density of people and infrastructure increases, potential inundation areas show a relative concentration of vulnerable people.

As well as raising socio-economic concerns, the use of levees has also elicited questions regarding changing hydrological conditions, environmental impacts throughout the watershed, losses of biodiversity, and structural concerns about maintenance and safety (Opperman et al., 2009). Liao (2014) shows that the presence of levees has hydraulic effects, such as increase of peak water levels, in his case study of the Lower Green River Valley in King County, Washington. Turning to overall risks, Merz, Elmer, and Thieken (2009) analyzed the contribution of 'high probability/low damage' floods versus 'low probability/high damage' floods to the expected annual damage from riverine floods in Germany, and concluded that expected annual damage is dominated by 'high probability/low damage' events.

While the above studies are empirical, the levee effect has recently emerged as a useful concept in conceptual modelling of the socio-hydrology of floodplains. These models use one of two approaches: agent-based modelling or dynamic systems modelling. In the first category, Reilly et al. (2017) couple individual behaviour on home adjustment with the level of protection offered by levees using, a county in Maryland (US) as example. They show how different behaviours affect the resulting societal vulnerability. Also in the US, this time illustrated by the North Dakota town of Fargo, Tonn and Guikema (2018) include not only home adjustments but also flood warning in their agent-based model, while Logan et al. (2018) focus on relocation to less flood-prone areas as a response to flood risk from tsunamis in Japan.

In the dynamic systems models, the levee effect figures as one of two dominant patterns of human–flood interactions (Di Baldassarre, Kooy, Kemerink, & Brandimarte, 2013a; Di Baldassarre et al., 2013b; Ferdous et al., 2018) (Figure 3.1). As in White's (1945) original definition, the levee effect results when flood protection structures lead to less frequent but more severe flooding through breaching or overtopping, which is often associated with increasing vulnerability (Di Baldassarre et al., 2015). Again, 'vulnerability' here means 'increased potential damage', consistent with White's (1945) hypothesis. The 'adaptation effect' refers to societies where the use of flood defence technology is limited, resulting in frequent flooding, which can be associated with decreasing vulnerability. Most floodplains combine both patterns, when only limited protection is provided, for example by the construction of submersible levees. To investigate these two patterns, Di Baldassarre et al. (2015) developed a conceptual mathematical model which is able to capture and explain the human–flood dynamics in two main prototypes of societies: green societies adapt to flooding by relocating out of hazardous areas; while technological societies construct levees to protect existing property.

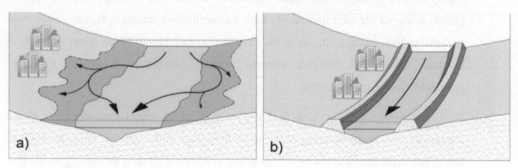

Figure 3.1: Schematic depiction of human adjustments to flooding: (a) adaptation: settling away from the river; (b) the levee effect: raising levees or dikes. (Source: Di Baldassarre et al., 2013b).

Ciullo, Viglione, Castellarin, Crisci, and Di Baldassarre (2017) used data from Bangladesh and Rome (Italy) to explore plausible trajectories of flood risk over time, assuming that Bangladesh is an example of a green society. While the present chapter shows this is not the full picture (because levees do play a role in Bangladesh), the generic results of their analysis provide a useful contribution to this exploration of the levee effect. Their main conclusion is that overall flood risk tends to be significantly lower in technological systems than in green systems. Technological systems may undergo catastrophic events, which lead to much higher ad hoc losses, but these add less to overall risk than the lower flood frequency. They assert that green systems are therefore more capable of withstanding environmental and social changes. However, in Bangladesh relocation is not usually possible, because land is in such short supply. The only alternative is to give up farming and migrate away to find employment elsewhere. In many other countries, too, relocation is a major challenge (Hino, Field, & Mach, 2017). It might encounter resistance or may not be economically or technically viable, as in the case of high-density high-investment coastal mega-cities, or locations that can only be inhabited thanks to flood protection, such as the Netherlands. Other non-structural measures can be considered to cope with flooding. They include early warning systems, flood proof building codes, and risk transfer (e.g., flood insurance). They applicability depends on the socio-hydrological context. Here we investigate how the levee effect in Bangladesh compares to these existing studies, focussing on flood risk, economic development and population density.

3.3 STUDY AREA

Bangladesh is one of the most flood-prone countries in the world. It is nevertheless a very densely populated country with more than 140 million people (964 per km^2). Around 80% of the population lives in floodplain areas (Tingsanchali & Karim, 2005) and depends on agriculture and fisheries (BBS, 2013). In most years, 25–30% of the floodplain area is inundated by the seasonal monsoon (Brammer, 2004). According to the classification by the Flood Forecasting and Warning Centre, who categorize flood events as normal, moderate and severe based on flood duration, exposure, depth and damage, recent severe flood events were observed in 1954, 1955, 1974, 1987, 1988, 1998, 2004, 2007 and 2017 (FFWC/BWDB, 2018). The maximum annual discharge of Jamuna River in the monsoon was measured at more than 100,000 m^3/s (RBIP, 2015). Successive governments have developed and implemented flood control measures to protect agriculture and populations from floods (Sultana, Johnson, & Thompson, 2008).

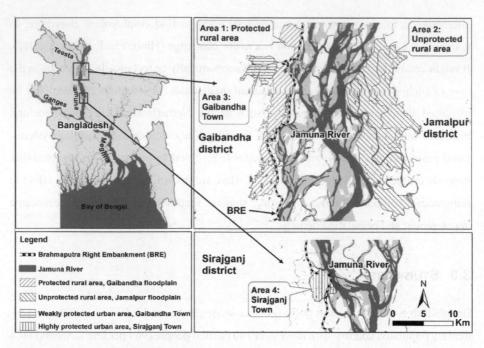

Figure 3.2: Bangladesh map with study areas.

One large intervention was the BRE, which was built in the 1960s to limit flooding in an area of about 240,000 ha on the western and southern sides of the Brahmaputra-Jamuna and Teesta Rivers, respectively (Areas 1, 3 and 4 in Figure 3.2). The aim was to increase agricultural production in that area. Construction of the BRE started in 1963 and was completed in 1968 at a cost of about BDT 80 million (~USD 16.7 million) (CEGIS, 2007). At present, the BRE has different characteristics along its length. It was constructed as an earthen embankment, which is what remains in Gaibandha District (Areas 1 and 3 in Figure 3.2; photo 1 in Figure 3.3). In this area maintenance since construction has been sporadic, and breaches are frequent in the monsoon due to the devastating force of the river. In Sirajganj District, the BRE has been heightened and reinforced with concrete since the 1990s, which is an ongoing process due to recurrent damage from floods (Area 4 in Figure 3.2; photo 3 in Figure 3.3). Breaches have occurred here occasionally, sometimes causing catastrophic damage (RBIP, 2015). On the left bank, no man-made levee was constructed (photo 2 in Figure 3.3). These differing situations (no levee, weak levee, strong levee), combined with two types of land use (urban and rural), provide the four areas (in two pairs) for our research (Figure 3.2). The four areas will be described in more detail shortly.

Figure 3.3: Current condition of the Brahmaputra Right Embankment at Gaibandha (areas 1 and 3) and Sirajganj (area 4) and natural levee at Jamalpur (area 2).

Ideally, to establish the levee effect in the study area we would need data on the areas' pre-BRE economic condition, that is, before 1970. However, the only historic economic data available pertain to the number of commercial establishments engaged in non-farming activities from 1971, and they present the data for whole districts, rather than the level at which we conduct our study (unions). An 'establishment' is defined as an enterprise or part of an enterprise that is situated in a single location and in which only a single (non-ancillary) productive activity is carried out or in which the principal productive activity accounts for most of the value added (BBS, 2016). These data do not give information about company size or its impact on the local economy. Some indication of these is available only for 2003 and 2013 as the number of employees engaged in these establishments (Table 3.1).

From these sparse data, it appears that shortly after the construction of the BRE the socio-economic situations in Gaibandha and Sirajganj Districts were similar, but in Jamalpur District it was less favourable (Table 3.1). From the number of establishments and the number of employees, it appears that in 2013 Sirajganj is best-off, with Jamalpur second and Gaibandha last. These numbers would support very generally what we expect to find regarding the levee effect in the two paired areas: Sirajganj has indeed seen more investment than Jamalpur, and also more than Gaibandha. These relative positions remain the same when the total number of employees is considered, with larger companies in Sirajganj than elsewhere. Larger companies generally require more investment, so this reinforces the observation that investors are prepared to invest more in the best-protected area, Sirajganj. In Jamalpur District the Jamuna Fertilizer Company, established in 1991, makes a huge contribution to the local economy, with about 6000 employees, somewhat skewing the data.

Table 3.1: Non-farm economic activities: number of establishments and employees.

Year	Gaibandha district, with weak man-made levee, Brahmaputra Right Embankment - includes areas 1 and 3			Jamalpur district, with no man-made levee - includes area 2			Sirajganj district, with strong man-made levee, Brahmaputra Right Embankment - includes area 4		
	Establishments	Persons engaged	Persons per establishment	Establishments	Persons engaged	Persons per establishment	Establishments	Persons engaged	Persons per establishment
1971	2675			1931			2392		
1989	10540			9329			11724		
1999	33651			32304			38226		
2003	62655	155094	2.48	54724	128366	2.35	84049	350363	4.17
2009	121495			127012			140029		
2013	151052	318579	2.11	159156	299997	1.89	174643	580698	3.33

Source: BBS (2016).

As we will discuss further in the Results section, our much more detailed data show several other differences in the socio-economic development in the study area. As with the district data in Table 3.1, these point at a real impact, or levee effect, of the BRE, although it is impossible to prove beyond doubt how much of the observed differences can be directly or indirectly attributed to the BRE and are not due to other factors, such as differential improvements in road access or investment opportunities. This is a chicken-and-egg situation: it cannot be asserted whether these developments are due to and/or independent of the BRE, and to what extent.

We now introduce the four areas where we investigated the existence of the levee effect by comparing two pairs of study areas:

- two rural areas: Area 1 (protected rural area, right bank of BRE in Gaibandha District) and Area 2 (unprotected rural area, left bank in Jamalpur District); and

- two urban areas on the right bank: Area 3 (Gaibandha Town, with weak BRE) and Area 4 (Sirajganj Town, with strong BRE)..

It is important to note that flooding still occurs in all four areas, though its frequency and magnitude varies, as explained below.

3.3.1 Area 1: protected rural area, right bank of BRE in Gaibandha District

This is a rural area in Gaibandha District, protected from normal floods (as defined by FFWC/BWDB, see above) by the BRE. Nevertheless, parts of this area are frequently inundated with excess rainwater due to its limited drainage capacity. Furthermore, a few locations of this area are also inundated by adjacent smaller rivers, the Ghagot and Alai. Also, sometimes the BRE breaches in this area, for example on 29 July 2016 (observed during the fieldwork for this research), when thousands of homesteads were submerged; the breach was not repaired until 2 November 2016 (observed during another field visit). The total area is about 74 km^2, with a population of approximately 111,000 (BBS, 2013).

3.3.2 Area 2: unprotected rural area, left bank in Jamalpur District

This area is on the left bank of the Jamuna River, with no man-made flood protection measures. Along the left bank flooding occurs more frequently than on the right bank, also due to other smaller rivers, like the Old Brahmaputra and Jinjira. High water in these small rivers sometimes occurs independently of high levels in the Jamuna, as they are not part of the same drainage basin. Riverbank erosion is prominent in this area. Flooding and riverbank erosion harm agriculture, homesteads and businesses, in turn impoverishing people. Migration is one of the main coping strategies, while several farm households also adapt their cropping patterns to accommodate flooding and cultivate fast-growing but less profitable crops after the flood season (Ferdous et al., 2018). The total area is about 174 km², with a population of approximately 148,000 (BBS, 2013).

The frequency of flooding in the protected and unprotected rural areas was not significantly different, as reported by the respondents to our survey. Figure 3.4 shows the percentage of households that remember a major flood in each year. This is not systematically higher in Area 2 than in Area 1. The total number of years that flooding was reported by any household was in fact slightly higher in Area 1 (34 years) than in Area 2 (32 years), but this difference is also not significant.

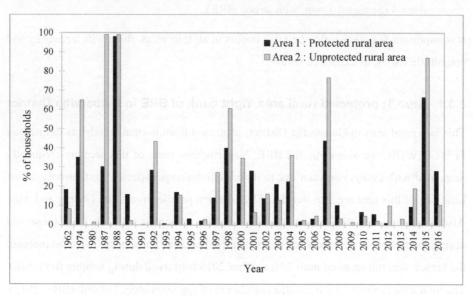

Figure 3.4: Major floods reported in rural areas, 1960–2016.

3.3.3   Area 3: Gaibandha Town, with weak levee (BRE)

This area is protected from 'normal floods' in the Jamuna River by the BRE. There are also two other embankments which prevent flooding from the small rivers Ghagot and Alai, tributaries of the Jamuna. These two embankments were constructed in 1995 by the Bangladesh Water Development Board. In this area, the last severe flood occurred in 1988. The BRE effectively protects the area against frequent flooding from the Jamuna, and as a result inhabitants feel relatively confident to invest in businesses and homesteads (Rahman, 2017). The total area is about 17 km^2, with a population of approximately 68,000 (BBS, 2013).

3.3.4 Area 4: Sirajganj Town, with strong levee (BRE)

This area is protected by the BRE, which was heightened and reinforced in the early 1990s to protect the town from flooding. The last severe flood occurred in 1988 (Sultana et al., 2008). The BRE effectively protects the area against most flooding from the Jamuna, so inhabitants feel relatively confident to invest in businesses and homesteads (Rahman, 2017). The total area is about 19 km^2, with a population of approximately 160,000 (BBS, 2013). The BRE at Sirajganj still breaches sometimes, but when this happens repairs gets more priority than in Gaibandha. As an example, the embankment just upstream of Sirajganj Town breached on 13 July 2017, and it was repaired within the next couple of days (observed during field visit).

3.4 RESEARCH APPROACH

The analysis of the levee effect in the study area in the Jamuna floodplain is based on primary data and secondary data. We summarize our data here; more details are provided in the Appendix D. In the two rural areas, we performed 560 household surveys from October 2015 to April 2016 and September to November 2016 in Areas 1 and 2. The questionnaire for the rural area survey is provided in the Appendix A. In the two urban areas, data from Rahman (2017), collected between October and November 2016, were used. They include 31 questionnaires in Area 3 and 34 in Area 4 and cover 80% of private investors in both towns. The questionnaire for the urban area survey is provided in the Appendix E. Secondary data include time series of the national census collected from the Bureau of Statistics of Bangladesh, and time series of satellite images of 30 m spatial resolution collected from CEGIS,

3.5 RESULTS

3.5.1 Comparison of development in protected rural area (Gaibandha District, Area 1) and unprotected rural area (Jamalpur District, Area 2)

We expected that average household income, wealth and population density would be higher in Area 1 than in Area 2. We also expected that transport infrastructure development would be faster, and flood damage lower, in Area 1 than in Area 2.

3.5.1.1 Flood damages

Flood damage information was available from primary data for severe events in 1987, 1988 and 2007 and the normal flood in 2015. In most years damages from flooding were less in Area 1 than in Area 2, except in the severe flood of 1988 (Figure 3.5).

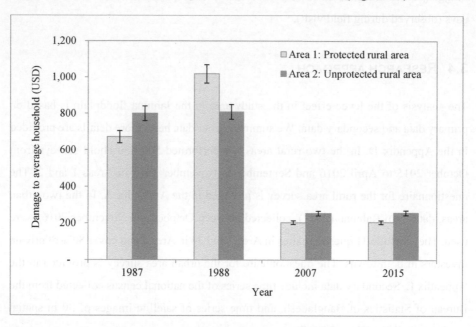

Figure 3.5: Comparison of average damages in different floods (error bar indicates 5% error margin of the median value).

3.5.1.2 Household income and wealth

About 80%of people in the study areas are farmers: their income and wealth mostly depend on their agricultural land and production. We calculated the total wealth of a household by summing up the value of all assets as indicated by the respondents; these include agriculture land, ponds, homestead land, other land (if any), housing materials, livestock and savings.

The current average wealth distribution shows clearly the economic differences between the households in Areas 1 and 2. In Area 1, people have more wealth (on average about USD 19,000) than in Area 2 in Jamalpur District (on average about USD 8,000). One could easily argue that the situation has always been like this, even before the construction of the BRE, since we do not have detailed data for the 1960s. To verify this, we collected qualitative information when doing the household surveys. The respondents in Area 1 agreed that their wealth and income status are much higher than before the BRE, partly because land values increased after the construction of BRE. As a result, the net wealth in Area 1 has increased more than in Area 2.

3.5.1.3 Expansion of settlement areas

The BRE has promoted general changes in land use from agriculture to settlement (housing), the latter requiring much higher investment and but also suffering higher damages in case of flooding. Figures 3.6 and 3.7 clearly show that in both locations settlement areas have increased, but expansion is faster in area 1 than area 2. In 1989 the settlement areas were almost the same, about 5% of the total area, but it increased to about 30% in the area 1 in 2014 whereas it is about 15% in the area 2 (Figure 3.6).

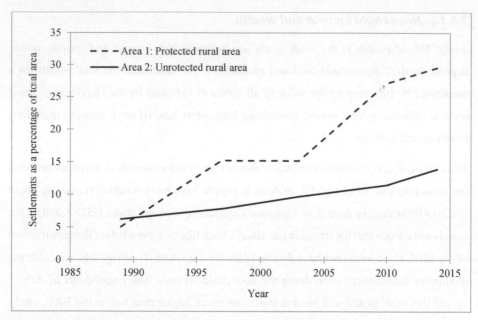

Figure 3.6: Expansion of rural settlement areas, 1989–2014.

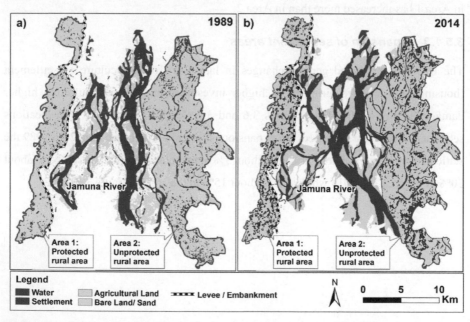

Figure 3.7: Expansion of rural settlement areas between 1989 and 2014.

3.5.1.4 Population density

Population density in the two areas differs substantially. Based on the census data, the density in Area 1 (1000–3000 persons per km^2) has always been higher than in Area 2 (100–2,000 persons per km^2) (Figures 3.8 and 3.9).

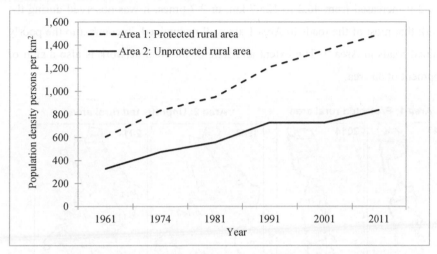

Figure 3.8: Increase in population density in rural areas.

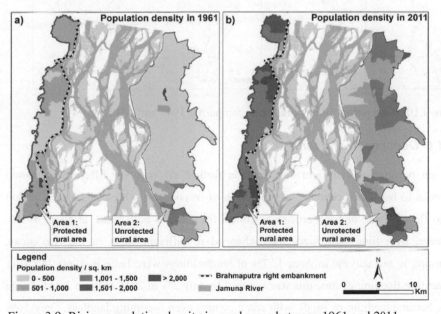

Figure 3.9: Rising population density in rural areas between 1961 and 2011.

61

3.5.1.5 Transport infrastructure

At the time of construction of the BRE, transport infrastructure in both areas was very similar, but since then it has been expanded much more in Area 1 (Figure 3.10). In Area 1, the total length of the road network increased from 15.4 to 145 km, or 9.4 times, while in Area 2 it increased from 46.2 to 125.7 km, or 2.7 times. It was observed during the fieldwork that most of the roads in Area 1 are well maintained, in contrast to the poorly maintained roads in Area 2. The extent and state of the road network is also a sign of development of an area.

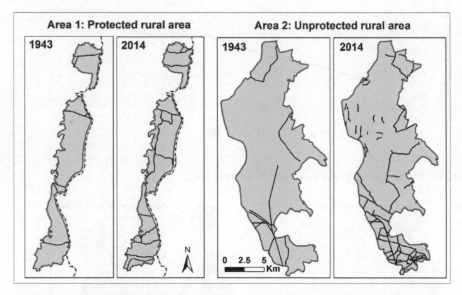

Figure 3.10: Road networks in rural areas, comparing 1943 and 2014.

3.5.1.6 Occupation and farm size

Data on occupation and farm size provide further evidence of differing economic responses to better protection. The proportion of respondents who changed occupation due to flooding and/or bank erosion is almost the same in both areas, at about 4–5%. However, over time farm sizes have changed in different ways in the two areas. According to our survey, in Area 1, 7% of landholdings were large in 1960, but after consecutive flooding events, this was reduced to only 2% in 2015 (Figure 3.11). Large farms became medium or small, with some owners even becoming landless. Only 16%

of households were landless in 1960, but 28% were in 2015. In Area 2 a comparable pattern can be observed, but there were more large farms to start with, and the decline was steeper. Many respondents from Area 2 mentioned that they could not recover from consecutive losses due to flooding and riverbank erosion. Many had to change their occupation temporarily, and 2% had changed their occupation permanently from farmer to day labourer. While the economic situation of farmers has worsened everywhere, the decline was more marked in Area 2.

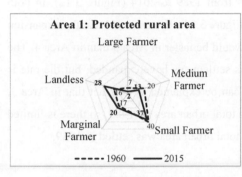
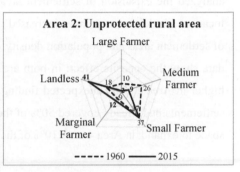

Farmer categories by Bangladesh government: *[1 acre = 0.405 hectare]*
- Large Farmer (land > 7.5 acre)   - Medium Farmer (land, 2.5 – 7.5 acre)
- Small Farmer (land, 0.5 – 2.5 acre)   - Marginal Farmer (land, 0.05 - 0.5 acre)
- Landless (land, 0 acre)

Figure 3.11: Agricultural land holdings in 1960 versus 2015 (percentage of households).

3.5.2 Comparison of urban development in areas with strong levee (Siranjganj Town, Area 3) and weak levee (Gaibandha Town, Area 4)

We expected that the expansion of settlement areas and population density would be faster in Area 4 than in Area 3, and the same for private investment and transport infrastructure development. To start our questionnaire, we asked respondents whether their investment decisions were affected by flood risk. Respondents indicated that neither Area 3 nor Area 4 had faced any flooding since 1988. In spite of several breaches in the BRE near Gaibandha, two other embankments nearer to Area 4 (on the right bank of the Ghagot and Alai Rivers) prevent floodwater reaching Area 4.

Thanks to this protection and the reinforced BRE in Area 3, respondents in both locations say they feel confident that they can invest. Another factor in investment decisions is the scarcity of land, which means that investors do not have much choice of where to invest: if they can get hold of any land, they will knowingly take the flood risk, expecting to recover their losses within a few years in case of flooding. We now investigate how this general attitude to flood risk is reflected our data.

3.5.2.1 Expansion of urban settlement areas

Due to the lack of remote sensing data with good spatial resolution prior to 1989, we analyzed the expansion of settlement areas from 1989 to 2014 (Figure 3.12). In both locations, settlement areas have increased (Figure 3.13). We expected that the expansion of settlement areas and population density would be faster in Area 3 than in Area 4. The data show the opposite effect: in both areas settlements have expanded, but the rate is higher in Area 4. This unexpected finding can be explained by the fact that in Area 3, settlements had already covered 50% of the total urban area by 1989, so there is limited space to expand; in Area 4 only 10% of the total urban area was settled in 1989.

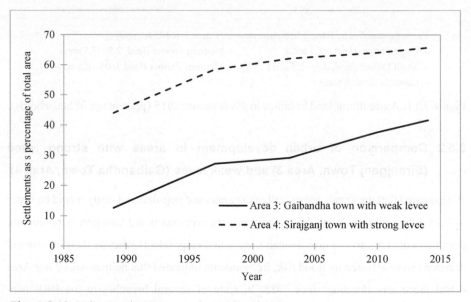

Figure 3.12: Urban settlement areas in 1989 and 2014.

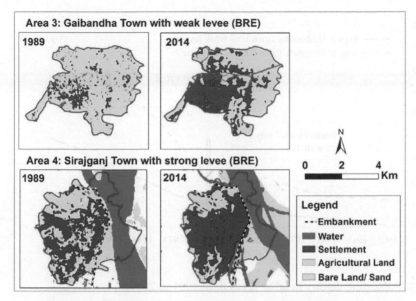

Figure 3.13: Expansion of urban settlement areas, 1989–2014.

3.5.2.2 Population density

Before the construction of the BRE, from 1901 to 1961, the rates of increase of population density in Areas 3 and 4 were similar (Figure 3.14), at 21 persons per km^2 per year in Area 3 and 16 persons per km^2 per year in Area 4. The rate raised dramatically after the construction of BRE in the 1960s, to 93 persons per km^2 per year in Area 3 and 44 persons per km^2 per year in Area 4 during 1961–1991. Since the reinforcement of the BRE near Area 3, the rate is still higher in Area 3 (158 persons per km^2 per year) than in Area 4 (80 persons per km^2 per year). These rates of change could of course be due to other factors besides flood risk, such as overall population growth rates, but nevertheless the difference in growth rates is significant in view of a levee effect.

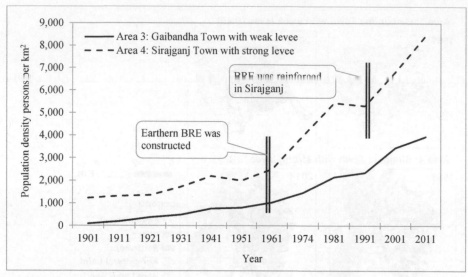

Figure 3.14: Population density in urban areas from 1901 to 2011.

3.5.2.3 Private investments

The main private investments are factories, non-governmental schools, hotels, fuel pump stations, housing for rent and a hospital (Figure 3.15). The differences are small, but investment in factories is slightly higher in Area 3, because most of the factories in Area 3 are textile mills and handloom factories, which need more capital than the small factories in Area 4. In the questionnaire, lack of capital for large investments, lack of technical persons, few marketing opportunities and migration to the capital city were listed as the main barriers.

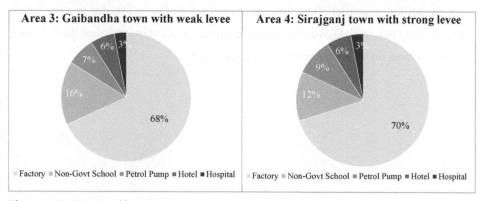

Figure 3.15: Types of investments in urban areas.

3.5.2.4 Transport infrastructure

At the time of construction of the BRE, transport infrastructure in both town areas was very similar, but since then it has increased 4 times (from 20.7 to 83.5 km) in Area 3 but only 1.8 times (from 20.6 to 37.5 km) in Area 4. In both towns, most of the roads are well maintained. The increase in road networks is clearly visible in Figure 3.16.

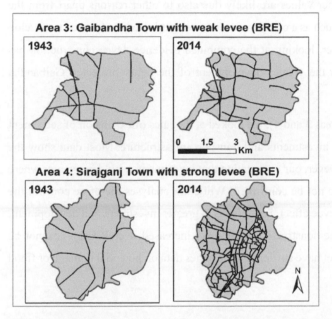

Figure 3.16: Road network in Gaibandha and Sirajganj Towns in 1943 and 2014.

3.6 DISCUSSION

We investigated whether a levee effect could be discerned in three districts along the Jamuna River by comparing historical socio-economic development and the current state of affairs using primary and secondary data. We compared four areas distinguished by different protection levels (good, moderate and none) and by dominant economic activity (rural or urban). Overall, our analysis shows how different levels of protection shape socio-economic development along the Jamuna River.

In the rural areas (Areas 1 and 2), data on average household income and wealth, expansion of settlement areas, population density, transport infrastructure, flood damage and occupations provide evidence of the greater investment due to levees postulated by White (1945). We confirmed that the flood losses experienced over time by residents were indeed lower in the protected area, but higher in severe events. We observed, as expected, higher average household income and wealth, greater expansion of settlement areas, and more rapid growth of population density and transport infrastructure in the protected area. These higher values are likely due also to other reasons apart from the level of flood protection, such as a concurrence between settlement and a need to develop the road network. However, looking at the combined evidence, levee construction has certainly played a role in the greater development of the better protected Gaibandha District.

In the two urban areas (Areas 3 and 4), we looked at the rates of expansion of settlement areas, population density, investments and transport infrastructure. Most data show the expected differences. However, our expectation that investments would be greater where the levee is stronger could not be confirmed. While our analyses therefore confirm the initial formulation of the levee effect, i.e., it leads to greater investment and development, the secondary effect of the construction of levees –increased overall risk – cannot be assessed, because we do not have sufficient time series data on potential damage or flood frequency.

3.7 CONCLUSIONS

We used an empirical approach to explore the levee effect postulated by White (1945) in four locations along the Jamuna River with different flood protection characteristics. The flood protection measures reduced flood frequency and enabled more urbanization and faster economic growth in urban areas, and higher incomes and lower losses in rural areas. In rural areas, the differences in economic development are statistically significant.

For Bangladesh as a whole, Mechler and Bouwer (2015) and Tanoue et al. (2016) present global analyses of the development of disaster losses due to flooding and their relation to vulnerability (defined here as propensity to incur losses, for which the protection offered by levees is taken as proxy). Both studies find that relative mortality (as a percentage of exposed population) and relative losses (as a percentage of GDP) have decreased over time in Bangladesh. Absolute casualties have decreased dramatically due to better flood protection, flood warning and evacuation, but absolute economic losses for the country as a whole have increased due to increased investment, in spite of the better protection offered by the levees. In our study area, the damages experienced by households have decreased since 1960 (Figure 3.17), which may be a result of adjustments to homesteads and cropping patterns, but it also indicates that people have less to lose, because they have fewer assets, due to repeated flooding.

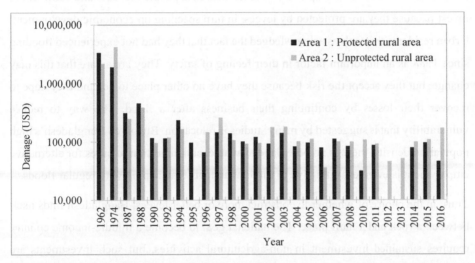

Figure 3.17: Total reported damage to households in rural areas (the vertical scale is logarithmic).

These findings, both for our small area and for Bangladesh as a whole, could therefore suggest that more and better levees should be constructed to reduce vulnerability and foster economic development. However, there are many reasons why this should not be concluded too soon. First, the levees offer limited protection due to their fragile state (in areas away from Siranjganj), so levee breaches are frequent. This situation can only be remedied by another round of capital investment to improve the current levees. In the 1990s, one stretch of 2.5 km cost USD 40 million, so to reinforce the whole BRE might run on the order of USD 3.5 billion. Also, maintaining a strong levee at Siranjganj requires frequent repairs, at around USD 10 million per year. These expenses are not matched by the expected economic returns. They would also mean that Bangladesh is forever locked into donor dependency if it decides to go down the route of higher flood

69

protection. Besides, there are doubts about the technical feasibility of civil engineering works to regulate the river – given its magnitude and sediment load, full control of flooding may not be possible (Rasid & Paul, 1987).

As long as levees breach frequently, greater investments in protected areas lead to more damages when the levee breaches, evidenced by our finding that during the highest floods the better protected rural areas experience higher damage than non-protected areas. In urban areas we found that despite the high residual risk, people feel safe and confident to invest because they are protected by levees, in turn speeding up economic development. Urban residents themselves acknowledged the fact that they had not experienced flooding since 1988 as an important factor in their feeling of safety. They are aware that this may change, but they accept the risk because they have no other place to go, and they hope to recover their losses by continuing their business after a flood. The way to reduce vulnerability that is suggested by many studies is relocation. But given Bangladesh's high population density, there is no other land available, and migration to cities for alternative employment is even less attractive to rural inhabitants than coping with regular floods.

Thus, in the choice of whether to construct more or better levees, Bangladesh finds itself between a rock and a hard place. The national goal to become a middle-income country requires sustained investment in non-agricultural activities, but such investments are threatened by high flood hazards. While we find that levees do contribute to such economic development, the construction and maintenance of levees along the Brahmaputra/Jamuna is very costly, leading to dependency on foreign investors and loans, and the levees are not even guaranteed to keep the enormous river under control. At the same time, there are limits to reducing vulnerability by other means. Flood warning and evacuation have saved many lives, but floods still lead to loss of income, land and other property. Relocation of rural populations to other rural areas is an option only for few communities, since very little spare land is available, and about half of Bangladesh's territory is flood-prone. Alternative employment in urban areas is so far limited and only a last resort due to poor housing and working conditions.

4

THE COSTS OF LIVING WITH FLOODS IN THE JAMUNA FLOODPLAIN IN BANGLADESH

This chapter is published as:

Ferdous, M.R.; Wesselink, A.; Brandimarte, L.; Slager, K.; Zwarteveen, M. and Di Baldassarre, G. (2019). The Costs of Living with Floods in the Jamuna Floodplain in Bangladesh. Water, 11, 1238, https://www.mdpi.com/2073-4441/11/6/1238.

Abstract

Bangladeshi people use multiple strategies to live with flooding events and associated riverbank erosion. They relocate, evacuate their homes temporarily, change cropping patterns, and supplement their income from migrating household members. In this way, they can reduce the negative impact of floods on their livelihoods. However, these societal responses also have negative outcomes, such as impoverishment. This research collects quantitative household data and analyzes changes of livelihood conditions over recent decades in a large floodplain area in north-west Bangladesh. It is found that while residents cope with flooding events, they do not achieve successful adaptation. With every flooding, people lose income and assets, which they can only partially recover. As such, they are getting poorer, and therefore less able to make structural adjustments that would allow adaptation in the longer term.

4.1 INTRODUCTION

The image of Bangladesh as a country that is adapting well stems from its long history of living with floods. Bangladesh is a riverine country and one of the most flood-prone countries in the world (Paul and Routray, 2010; Roy et al., 2016; Mamun, 2016). The country is in the largest delta of the world, which developed, and is continuously changing, through processes of sedimentation and erosion by the three mighty rivers Ganges, Brahmaputra, and Meghna. Bangladesh is also a very densely populated country with most people depending on agriculture and fisheries for their livelihoods (BBS, 2013). Most of the population lives in floodplain areas, with varying degrees of exposure to riverine flooding and riverbank erosion (Tingsanchali and Karim, 2005). In Bangladesh, floods can be both resources and hazards. In the monsoon season, 25–30% of the floodplain area is typically inundated by the so-called normal floods (*borsha*). This is considered beneficial as flooding increases the fertility of the land by depositing fresh silt on the soil and replenishing soil moisture without doing too much damage to property or disruption to traffic and commerce. However, an abnormal flood (*bonna*) is considered a hazard because of the damage it does to crops and properties, and the threat to human lives (Paul, 1984; Brammer, 1990). Riverbank erosion is another hazard in times of flood and occurs almost in every year, whether the flood is normal or abnormal. Abnormal floods and riverbank erosion not only cause substantial damage to inhabitants' agricultural lands, crops, and properties, they also have wider socio-economic consequences, for example a decline in agricultural production (Banerjee, 2011) and a steady flow of migrants to urban areas (Gray and Mueller, 2012; Joarder and Miller, 2013; Fenton et al., 2017). Each year several thousands of people become temporarily or permanently homeless and/or landless and take refuge to nearby embankments or on neighbors' or relatives' land. The extremely poor people who live on the islands in the wide (up to 16 km) rivers (called *chars*) are most exposed to and affected by flood hazards and riverbank erosion. In the 1988 flood, more than 45 million people were displaced and over 3000 died (Haque and Zaman, 1993; Zaman, 1993). Flood mortality rates have declined since 1988 because of better institutional support during floods (Khandker, 2007). Yet, the socio-economic disruption caused by flooding events is still considerable (Brouwer et al., 2007; Paul and Routray, 2010).

"Living in a low-lying and densely populated country on the front line of climate change, Bangladeshis are taking a lead in adapting to rising temperatures and campaigning to limit climate change. Bangladeshis will keep their heads above water, but at huge costs" (Roy et al., 2016). This back cover summary of the recent book *"Bangladesh Confronts Climate Change: Keeping Our Heads Above Water"* (Roy et al., 2016) portrays a common, positive image of Bangladesh as being able to respond successfully, or adapt, to climate change (see also Haque and Zaman, 1993; Paul, 1997; Brammer, 2004; Di Baldassarre et al., 2015; Ciullo et al., 2017). However, it also indicates that this response comes with huge costs. Floodplain inhabitants use a wide range of tried-and-tested strategies to cope with the conditions during and after the flooding. However, repeated exposure to flooding events often means impoverishment.

Only a few authors have paid sufficient attention to this effect, which can be seen as a sign of maladaptation (Barnett and O'Neill, 2010). Most research done on people's abilities to cope or adapt is based on qualitative statements from floodplain inhabitants, without looking at the cumulative effect of floods over a longer time period (i.e., across decades). As time series of quantitative data to substantiate these claims are not available from secondary sources, estimates of flood damage often rely on unverifiable proxies (Yang et al., 2015). More specifically, various studies assert that people are getting poorer because of recurring floods (Alamgir, 1980; Rahman, 1989; Haque and Zaman, 1993), but most of these do not provide quantitative substantiation. One of the processes that may lead to impoverishment is diminished agricultural production in times of disastrous floods (Banerjee, 2011). Rural people are most affected by these floods; they are also suffering from the persistent effects of labor market disruption and income deficiency (Sultana and Rayhan, 2012). Moreover, they must spend a large portion of their income on food and repairing or constructing new houses, as their old houses are often ruined by floods and riverbank erosion. Their savings are reduced to zero and the other necessary expenses to survive are only increasing their burden (Yasmin and Ahmed, 2013).

A few studies quantify the average household losses due to severe floods and the associated riverbank erosion. Chowdhury (1988) presents a figure of 47 and 88 USD/year for the years 1986-1987, in two locations. Ten years later Thompson and Tod (1998) estimate losses to homesteads to be 132 and 190 USD/year for the 1988 and 1991 floods, again in two locations. The most complete figures are given by Paul and Routray (2010), who estimate income losses at 82 and 108 USD/year and asset losses 191 and 257 during the 2005 flood, again in two locations. Combined with direct loss of assets in floods, and loss of the key productive asset (land) to erosion, severe floods in the chars make vulnerable households and the community poorer (Thompson and Tod, 1998). Finally, they are falling into debt and impoverishment (Alamgir, 1980; Rahman, 1989; Haque and Zaman, 1993; Hutton and Haque, 2004). Further indebtedness leads to even more drastic regimes of "eating simpler food", increased malnutrition, increased levels of disease, increased numbers of wives left to feed children alone while their husbands are temporarily migrated for work, increased sales of remaining assets, increased rural landlessness, increased migration to the cities, and increased vulnerability to future floods.

Moreover, only a few authors investigate the distributional effects of flood impacts. Paul and Routray (2010) show how the adoption of a particular set of adjustment strategies depends on people's socio-economic circumstances, such as education, income, and occupation. Islam et al., (2013) look at adjustment strategies against flood and riverbank erosion of the char inhabitants of the Jamuna River. They also find that household's ability to cope with flood and river erosion depends on people's socio-economic and environmental conditions. Hutton and Haque (2004) assert that impoverishment and marginalization in part reflect inequitable access to land and other resources. The likelihood of impoverishment of the household is further increased not only by social and demographic factors (including gender, education, health and age), but also by underlying economic and social relationships which can increase human vulnerability to risk. Indra (2000) investigates the forced migration due to riverbank erosion (and hence loss of land) in the Jamuna floodplain, and concludes that most people displaced by riverbank erosion are already poor and disempowered before being uprooted by the shifting channels of the Jamuna River.

This chapter critically assesses the idea that Bangladeshis have a good adaptive capacity to changing environmental conditions. To this end, it examines the socio-economic effects of repeated floods and riverbank erosion on rural households, and assesses whether and how different households deal with this flooding. The aim is to uncover whether and how households succeed in adapting to flood conditions, as opposed to those who are merely coping. As defined by Berman et al., (2012) coping refers to immediate responses to events, while adapting prepares households for expected future events. These definitions are extended as follows: coping includes long-term adjustments where the net result can be a decline in socio-economic conditions, while adapting means that people continue to live as well, or better than before, and are as well, or better than before, able to cope with future events. Both coping and adapting require the mobilization of a variety of technologies and social or institutional changes, which is defined as adjustments. The actions and strategies for adjustment may be the same, but have a different result: coping or adapting. In general, coping strategies are those that are possible within the current settings, and adaptation strategies are more likely to involve more fundamental changes in the type of livelihood activity or location (Rahman et al., 2015).

4.2 CASE STUDY

To understand how Bangladeshi cope with flooding, this study focuses on the floodplain area along a 30 km reach of the Jamuna River in the north of Bangladesh. The study area includes parts of Gaibandha district and parts of Jamalpur district (Figure 4.1). The total surface is about 500 km^2 and the total population is approximately 0.36 million people (BBS, 2013; BBS, 2014). This is a unique case study since three adjacent areas are inhabited by people with similar socio-economic conditions, but different levels of flood protection and exposure to riverbank erosion (Figure 4.1).

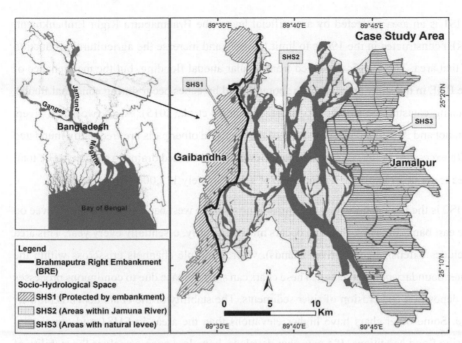

Figure 4.1: Bangladesh map with study area (Source: Ferdous et al., 2018).

To investigate the effect of floods and riverbank erosion on livelihood conditions, the area is divided into three socio-hydrological spaces (Ferdous et al., 2018). The socio-economic situations are also classified based on income levels and assets (Section 4.2.2)

4.2.1 Socio-hydrological spaces

Figure 4.1 shows the braided riverbed, which includes many inhabited river islands (*chars*) that get flooded with varying frequency (some every year, some only with severe floods). The area to the west of the river is protected by a human-made embankment, while the area to the east of the river only has a natural levee deposited by the river. These different physical conditions create different flood protection levels, which in turn give rise to different socio-economic responses and conditions. To reflect this socio-spatial differentiation, the authors have elsewhere explained and motivated the classification of the study area into three socio-hydrological spaces, abbreviated to socio-hydrological spaces (SHS) (Ferdous et al., 2018) (Figure 4.1). These spaces are briefly described below.

SHS1 is an area protected by an artificial levee (the Brahmaputra Right Embankment, BRE) constructed in the 1960s to limit flooding and increase the agricultural production of that area. This area is protected from regular annual flooding, but the maintenance of the BRE in the study area has been sporadic, and breaches occur during abnormal floods, causing catastrophic floods and damages (Sarker et al., 2015). Also, two small rivers Ghahot and Alai inundate some parts of the area, and other parts are frequently inundated with excess rainwater, due to their low elevation and limited drainage capacity. The total area is about 74 km^2 with a population of approximately 111,000 (BBS, 2013).

SHS2 is the floodplain within the embankment on the west bank and the natural levee on the east bank. In SHS2, flooding occurs most frequently, essentially every year. This area includes extensive chars (river islands), where multiple channels crisscross within the outer boundary of the riverbed. These chars can shift in space due to continuous processes of deposition and erosion of river sediments. The stability of the chars depends on their age. Some older chars have higher elevations than the areas in SHS1 and remain dry during flood conditions. If a new char develops, homeless people analyze the stability of the new char and after 2–3 years start activities such as farming or living on the char. When the age of the char exceeds about 20 years it is called a stable char (Sarker et al., 2003). Nevertheless, all chars can shift during the flooding events. The total area is about 246 km^2 with a population of approximately 104,000 (BBS, 2013).

SHS3 is the area on the east bank which is without any man-made flood protection. Flooding occurs here more frequently than SHS1, around once in two years (Ferdous et al., 2018). This area is sometimes flooded by adjacent small rivers, in this case the Old Brahmaputra and Jinjira, as well as by excess rainfall. Riverbank erosion is also prominent in this area. Inhabitants take the initiative to build small spurs and bank protection made from bamboo and wood, to try to stop erosion. However, while these encourage sedimentation at a local scale, they are not sufficient to stop large scale erosion. The total area is about 174 km^2 with a population of approximately 146,000 (BBS, 2013; BBS, 2014).

Decadal data from 1961 to 2011 show that the population density has increased from 300 persons/km^2 to about 800 persons/km^2 over the whole study area (BBS, 2013; BBS, 2014).

However, population density in the three spaces differs substantially. In SHS1 it has always been higher than in the SHS2 and SHS3 (Figure 4.2). The density in SHS1 has increased from 600 to 1500 person/km^2, a rate of 18 persons/km^2/year. In SHS2 population density has increased from 200 to 400 person/km^2, a rate of 4 persons/km^2/year, and in SHS3 has increased from 300 to 800 person/km^2, a rate of 10 persons/km^2/year.

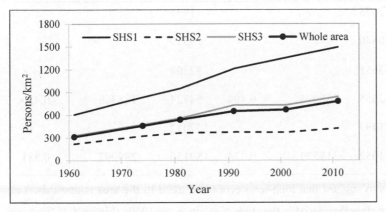

Figure 4.2: Population density of the study area in the period 1961–2011.

4.2.2 Socio-economic conditions

Data on socio-economic activities are sparse. The only historic economic data (from 1971) pertain to the number of commercial establishments developed for non-farming activities. These data are aggregated by districts that are larger than the SHS considered here. An establishment is defined an enterprise or part of an enterprise that is situated in a single location and in which only a single (non-ancillary) productive activity is carried out or in which the principal productive activity accounts for most of the value added (BBS, 2016a; BBS, 2016b). Table 4.1 shows that the socio-economic situation in Gaibandha district was a little better than in Jamalpur district in 1971, but they were almost the same in 2013. The growth in the number of establishments changed dramatically around the year 2000 (Table 4.1). To get information about the company size and its impact on the local economy, this research uses the number of employees as a proxy. This information is only available for 2003 and 2013 (Table 4.1).

Table 4.1: Non-farm economic activities: number of establishments and employees (BBS, 2016a; BBS, 2016b).

Year	Gaibandha district			Jamalpur district		
	Establishments	Persons engaged	persons/establishment	Establishments	Persons engaged	persons/establishment
1971	2675			1931		
1989	10540			9329		
1999	33651			32304		
2003	62655	155094	0.404	54724	128366	0.426
2009	121495			127012		
2013	151052	318579	0.474	159156	299997	0.531

Table 4.1 seems to suggest that long-term economic trend in the area is upwards. Yet, agricultural farm sizes decreased in the study area since the 1960s (Figure 4.3). This is a potential sign of impoverishment. Figure 4.3 shows that the number of large farms decreased from 10% to only 1% during the period 1960s to 2016 and the number of landless households increased from 10% to about 37% during the same period.

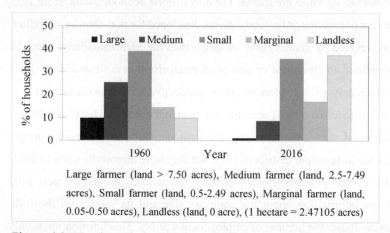

Large farmer (land > 7.50 acres), Medium farmer (land, 2.5-7.49 acres), Small farmer (land, 0.5-2.49 acres), Marginal farmer (land, 0.05-0.50 acres), Landless (land, 0 acre), (1 hectare = 2.47105 acres)

Figure 4.3: Number of non-farming establishments and change in agricultural farms in the study area.

In the study area, more than 80% of households rely principally on farm incomes, so agricultural income and assets are key indicators for socio-economic status. The existing Bangladesh Government classification of farm sizes only accounts for land holding size (BBS, 2018a). However, households have other assets (homesteads, equipment, cash or jewelry) that help them survive in case of flooding or riverbank erosion. Moreover, differences in average income affect how households can cope with such events. Data of this research include current annual income, annual expenditure and total wealth (see Section 4.3 for details about data collection). Using these data, the surveyed households are divided into five socio-economic classes that combine current wealth and income (Table 4.2). To increase the statistical significance of comparisons between classes, the class boundaries are chosen in such a way that each class contains an almost equal number of households.

Table 4.2: Classification of socio-economic status of the people in the study area.

Socio-economic classes	Total wealth and yearly income
Poor	Little wealth ≤ 5000 USD, Low income ≤ 750 USD/year
Moderate poor	Little wealth ≤ 5000 USD, Moderate income > 750 USD/year
Moderate	Moderate wealth 5,000 – 30,000 USD, Moderate income ≤ 1,000 USD/year
Moderate rich	Moderate wealth 5,000 – 30,000 USD, Moderate income > 1,000 USD/year
Rich	High wealth > 30,000 USD, Moderate income > 1,000 USD/year

By comparing the current distribution of socio-economic classes in the three SHS, see Section 4.2.1 and Figure 4.1), one can see remarkable differences (Figure 4.4a). In SHS1, 10% of the households are poor and 35% are rich, while in SHS2 it is just opposite. SHS3 takes an intermediate position, with 15% poor households and 25% rich. Using the Government farm size classification for comparison (Figure 4b), it is observed that more than 35% of households are landless in the study area and only 1% of households are large (Figure 4.4b). Most of the landless farmers are found in the SHS2 as they are experiencing much more land erosion than SHS1 and SHS3.

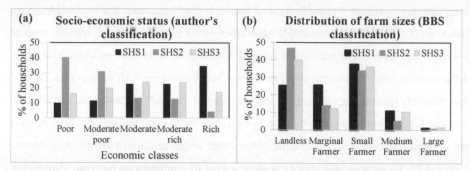

Figure 4.4: Current socio-economic status of the households in the study area by socio-hydrological space, (a) Socio-economic status (author's classification), (b) Distribution of farm sizes (BBS classification).

4.2.3 Adjustment strategies in the study area

Inhabitants of the study area employ a range of strategies to adjust with flooding. These can be categorized into those that allow survival in times of flood and those that allow long-term living with floods and riverbank erosion.

To survive during flooding, people eat fewer meals, borrow money or take a loan, or sell their labor cheaply in advance (Yasmin and Ahmed, 2013). If necessary they also sell their land, livestock, housing materials and other personal belongings, including jewelry and household goods (Thompson and Tod, 1998). In char areas (SHS2, Figure 4.1) the inhabitants only leave their homes when their lives are at risk. When high floods erode their land, they dismantle their houses and transport them to another char which is less (or not) affected by flooding and erosion (Schmuck, 2000). If necessary, people take shelter on the embankments, with their livestock, in the hope that they might return in the near future to the re-emerged land, where they have property rights (Brouwer et al., 2007). In most cases, these hopes are not fulfilled because it may take decades for land to re-appear (Haque, 1988). Others move to a nearby relative's or friend's house or migrate temporarily to other districts looking for temporary work (Shaw, 1989; Indra, 2000; Gray and Mueller, 2012; Findlay, 2012; Rahman, 2013; Ayeb-Karlsson et al., 2016). It is noted that main roads and railways are built on embankments to raise them above high flood levels. Rural roads and paths between settlements generally follow the highest land available and are usually also built on embankments to raise them above normal flood levels. This somewhat limits disruptions during floods, and provide an emergency refuge.

To structurally improve their chances of survival during a flood, inhabitants adjust their homes. In the char areas (SHS2), houses are built in a way that they can be easily dismantled (Shaw, 1989; Indra, 2000). In SHS1 and SHS3, homesteads are generally built on natural elevations, or on artificial earthen mounds, the height of which is determined by local experience of previous high flood levels (Brammer, 2004). The plinth of houses is further raised by digging earth from local depressions. Especially on the chars (SHS2) people also build platforms inside their homes to take shelter using bamboo, straw, water hyacinth, and banana stalks during abnormal flooding years (Haque and Zaman, 1993).

To improve livelihoods, rural inhabitants have developed farming practices that are adjusted to the height, duration, and timing of normal floods, i.e. that commence and recede in time and attain normal height (Paul, 1984; Ayeb-Karlsson et al., 2016). Rasid and Mallik (1995) observe that the most common adjustment of rice cropping to the uncertainties of the flood regime is evident from the practice of intercropping two rice varieties, aus and aman, together. This measure ensures that at least flood-tolerant aman would be secured during an abnormal flood regime, even if the flood-vulnerable aus is lost or damaged. During a normal flood regime both aus and aman would succeed, often resulting in a bumper crop. Farmers have made a careful selection of the best adjusted varieties of rice over the centuries, to enable them to face floods (Paul, 1997), and they select other crops to sow off-season to fit the land elevation (Haque and Zaman, 1993). Other adjustment strategies include early harvesting of aus in case of excess or early flooding, planting older and taller seedlings on lands liable to repeated flooding, re-transplanting salvaged seedlings, protection of rice plants from water hyacinth by floating bamboo fences, and the post-flood cultivation of lentils, pulses, mustard, as well as winter vegetables and wheat (Rasid and Mallik, 1995).

While temporary migration can be a survival strategy (see above), migration can also be a long-term adjustment to increase household incomes and livelihood security (Ayeb-

Karlsson et al., 2016). Very few people move permanently to towns and cities (Haque, 1988; Hutton and Haque, 2004; Joarder and Miller, 2013); the majority of flood affected people try to stay near their homes because kinship ties mean that local inhabitants help each other in crisis situations (Mamun, 2016). In that case, migration often concerns one or more family members who migrate, usually men. This can be on a seasonal basis, to other rural areas as agricultural laborer, or to urban areas as unskilled laborer or rickshaw driver, or more permanently to work in factories. Permanent migration out of the area of origin is rare, and mostly related to the loss of land due to riverbank erosion, while loss of livestock and crop failure more likely leads to temporary migration (Indra, 2000; Joarder and Miller, 2013). However, migration often comes at a cost: working conditions tend to be challenging and dangerous. People who are forced to migrate permanently from rural to urban areas often end up in slums with difficult living conditions.

4.3 DATA AND METHODS

Primary data and secondary data were collected to explore how households in the study area adjust to regular flooding and riverbank erosion, as well as the long-term socio-economic effects of these adjustments. We summarize our data here; more details are provided in the Appendix F.

Household surveys and focus group discussions (participants selected from the previously surveyed households) were performed during the dry seasons of 2015 and 2016. The principal set of primary data consists of 863 questionnaires dealing with several themes: general information (location of settlement and agricultural land, main occupation, age, income and expenditures, wealth and origin of the households), information on different flood experiences (depth of floods, frequency, duration, flood damages, effects on agricultural income and expenditures, other adjustment options such as migration) and experiences with river erosion (frequency, damages, migration, adjustment options etc.). For several of these aspects the respondents were asked to compile a historical record going back to 1960, which is why the households headed by older men or (occasionally) women were selected. This is a limitation of the study. Many old men were surveyed since they experienced many flooding events. Although other members of the family were typically present during the surveys, this choice may have introduced a selection bias. 12 focus group discussions were also set up in the study area to validate and contextualize survey data.

Since the respondents were asked to recollect their flood experiences going back to 1960, inaccuracies due to memory loss is one of the limitations of the household survey data. When asked to recollect major flood events and details of flood damages, respondents could easily remember events in the last two decades, but they were not so confident about the floods before 1980s. To handle this issue, the questionnaires were filled in in the presence of other family members, who helped the primary respondent to remember details about the past. The surveyors also used references to remarkable years, such as the year of independence of Bangladesh in 1971, or the construction period of embankments, to connect the respondents with years of major flood events.

To further check the reliability of the respondent's flood memory, the surveyed data are compared with both Government data and published journal articles. Based on flood duration, exposure, depth, and damage, the Flood Forecasting and Warning Centre (FFWC) of Bangladesh Water Development Board (BWDB) classifies flood events into three categories: normal, moderate, and severe. According to this classification, Bangladesh has experienced 9 severe floods since 1950s, namely in 1954, 1955, 1974, 1987, 1988, 1998, 2004, 2007 and 2017 (FFWC/BWDB, 2018). However, according to household surveys a total of 33 major floods were experienced in the study area since the 1960s. A plausible reason for this difference is that the study area is very close to the Jamuna River, so households are facing huge damages almost every year. As such, almost all floods have severe consequences in the study area.

In the literature different lists of severe floods are presented. Hossain et al. (1987) report major flood events in 1954, 1955, 1956, 1962, 1963, 1968, 1970, 1971, 1974, and 1984. Brammer (2004) mentions that 16 major floods struck Bangladesh in the years between 1950 and 2000, with additional floods in 1977, 1980, 1987, 1988, 1998, and 2000. The respondents in the study could easily remember those flood events mentioned by Brammer (2004) except the flood of 1977. According to Paul (1997), Bangladesh has experienced 28 major floods in the past 42 years (1954–1996), of which 11 were classified as "devastating" and five as "most devastating". Summarizing all published flood information in the journal articles, 30 major floods were observed during 1962–2017. This flood count is very close to the numbers that the study collected from the household surveys. This suggest that the memory of the respondents is good enough to compile a record of historical flood events in the study area.

For the data on land loss due to riverbank erosion, all respondents were confident that they could remember the actual losses and the years they occurred. According to them, the loss of land is never forgotten. Respondents claimed to remember very well how much lands they had in 1960s, how much land had eroded since then, and in which year. They said that it is possible to recover from flooding, but that it is not possible to recover losses from riverbank erosion.

4.4 RESULTS

Statistical analyses are performed for the whole study area, for each SHS (Figure 4.1), and for the different socio-economic classes (Table 4.2). This is done through statistical analysis of single variables and correlation between variables. ANOVA test ($p < 0.05$) and Chi-square tests are also performed to verify the significance of the outcomes. Statistical test summaries and supplimentary analysis for costs of living with floods are given in the Appendix G and Appendix H respectively.

4.4.1 Adjustment strategies

Relocation means to settle permanently in another place with the whole household. Households in the study area do not normally relocate due to flooding events. In case of severe flooding, some people may temporally evacuate over a short distance, to return to their houses after the flood. When households were asked about permanent relocation, many of them have expressed that "*it is very hard to live here but we are born here, our forefather lived here then why should we leave?*" They relocate permanently only when they face riverbank erosion. About 49% of households relocated during the entire period 1962–2016, and more than 95% of households claimed that riverbank erosion is the main reason of relocation. As expected, the maximum number of relocations occurred in the char areas, i.e., SHS2 (Figure 4.5a). By considering the socio-economic status, the maximum number of relocations occurred in poor (76%) and moderately poor (70%) households (Figure 4.5b). More than 90% of respondents relocated within 5 km from their previous locations and about 40% poor and 30% moderately poor households are considering relocating again because of riverbank erosion. A statistical analysis was performed and it was found that there exists a significant difference in relocation by the respondents between SHS and socio-economic groups (with α = 0.05).

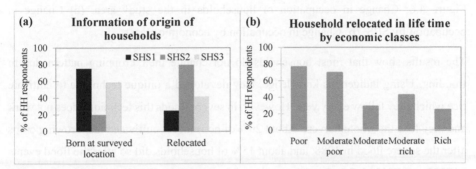

Figure 4.5: Households relocated due to riverbank erosion, (a) Information of origin of households, (b) Household relocated in lifetime by economic classes.

Only 5% of households mentioned that members had to change their occupation due to flood (Figure 4.6a). They were mainly from the poor and moderately poor classes (Figure 4.6b). They changed their occupation mainly from farmer to day laborer, rickshaw puller, and fishermen. The number of households reporting a change in occupation is low because alternative employment opportunities are limited. More than 90% of households mentioned that they have nothing to do during flood events and a scarcity of work arises during those periods. Normally, they look for day laborer work during flood events. A statistical analysis was performed and it showed that there is no significant difference in change in employment by the respondents between the SHS but there exists a significant difference in change in employment by socio-economic groups (with α = 0.05).

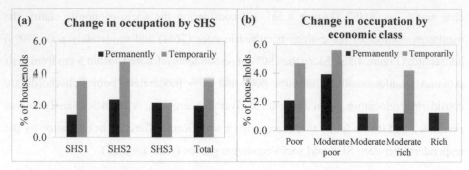

Figure 4.6: Change in occupation of households in the study area, (a) Change in occupation by SHS, (b) Change in occupation by economic class.

The results show that most households do not change their cropping pattern due to flooding. Using indigenous knowledge, they developed a unique technique to cultivate rice which they follow every year. However, in severe floods this technique does not work and they lose the whole harvest. About 20% of households cultivated fast growing crops after the severe flood in 1988, and about 15% of households did so after the flood events in 2007 and 2015. A few farmers mentioned that they keep their lands fallow during the flood season to avoid losses.

In the study area, inhabitants raise the plinth of their houses to prevent flood water to get into their home. Only 11% of household respondents have also raised their homestead platform. They all are from rich and moderately rich socio-economic classes.

Inhabitants of the study area are undertaking some other adjustment strategies such as storing food for flood event, constructing houses with movable materials, planting trees to avoid erosion. A few of them are also temporarily going to big cities to earn some money as day laborers to overcome the flood losses.

4.4.2 Impacts of flooding and riverbank erosion

By analyzing the negative impacts of flooding and riverbank erosion, it was found that the average income is decreasing over time. About 98% of household experienced decrease in income in every year. Households claim to lose about 90% of their monthly income during severe flood events (Table 4.3). A statistical analysis was performed and it showed that there is no significant difference in income loss of the respondents between the SHS and by socio-economic groups (with $\alpha = 0.05$).

Table 4.3: Decrease of income due to floods in the study area.

Parameters	Socio-hydrological spaces			Socio-economic classes					Whole study area
	SHS1	SHS2	SHS3	Poor	Moderate poor	Moderate	Moderate rich	Rich	
% of households experiencing a decrease of income	97%	99%	96%	97%	98%	97%	97%	99%	98%
% of monthly income lost during severe floods	86%	88%	91%	88%	88%	89%	88%	90%	89%

Moreover, 71% of households face an increase in monthly expenses in times of floods (Table 4.4). Such an increase is about 60% of the monthly income with little (but not significant, see Appendix G) differences between SHS and by socio-economic groups (Table 4.4). Decrease in income loss and increase in monthly expenses due to flood and riverbank erosion are generating the indebted situation for the households in the long run.

Table 4.4: Increase in expenses due to flood in the study area.

Parameters	Socio-hydrological spaces			Socio-economic classes					Whole study area
	SHS1	SHS2	SHS3	Poor	Moderate poor	Moderate	Moderate rich	Rich	
% of households facing an increase in expenses in time of flood	71%	77%	66%	78%	67%	69%	71%	70%	71%
% of monthly income of the average increase of expenses	63%	62%	60%	66%	50%	61%	52%	78%	62%

4.4.3 Impoverishment

Land loss caused by river erosion is the major loss for the inhabitants of the study area. Land loss data were used to better analyze whether impoverishment differs by SHS and socio-economic classes. On average, 0.9 ha of land per household was lost between 1962 and 2016. As expected, the inhabitants of SHS2 had the highest losses with 1.4 ha per household (Figure 4.7a). About 10% of households lost all their land since 1962 and 20% of households lost more than 80% of their land. Analyzed by socio-economic status, it is found that the poor households have lost the most, i.e., 1.5 ha per household (Figure 4.7b). Moreover, poor and moderately poor households lost the highest proportion of their land, many ending up with no land at all (Figure 4.7c,d). The households of SHS2 have lost about 80% of their lands during this period. A statistical analysis was performed and it showed that these differences in land loss between the SHS and socio-economic groups are significant (with α = 0.05).

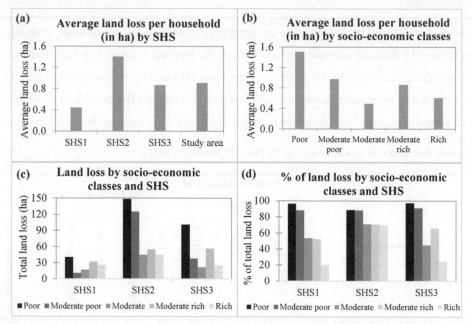

Figure 4.7: Total land loss by SHS and economic classes in the study area from 1962 to 2016, (a) Average land loss per household (in ha) by SHS, (b) Average land loss per

household (in ha) by socio-economic classes, (c) Land loss by socio-economic classes and SHS, (d) % of land loss by socio-economic classes and SHS.

The survey data also include time series on agricultural crop loss, homestead loss, and other asset losses caused by flooding and riverbank erosion. In terms of total asset losses, households in SHS3 have lost the most, around 2000 USD per year, while the households in SHS2 have lost the least, around 1250 USD per year. Households in SHS2 are better prepared, since they live in the char area, and have less to lose because they are, on average, poorer. By analyzing asset losses by socio-economic status, one can see that poor households have lost most assets (more than 2000 USD). Most poor people today claimed to be richer back in the 1960s, but that they became poorer because of repeated asset losses caused by flooding the riverine erosion. Yet, it should be mentioned that no significant correlations between flood severities and reported losses were found (details are given in Appendix H).

The respondents were asked about their recovery processes to explore how they recover from losses caused by consecutive events. *"Flood and riverbank erosion have snatched everything from us, and we become destitute"*, is one of the most commonly heard expressions in the char area (SHS2). People in the areas also like to ease their situation by believing that *"Almighty Allah (God) has sent this flood towards us and so we have to accept it. This is our fate"*. Most respondents informed us that it is almost impossible to recover from the riverbank erosion. About 53% of the respondents mentioned that they could not recover at all from the combined losses of flooding and riverbank erosion (Figure 4.8a). Only 5% of the households recovered a little but not enough to get back to their previous position. They recovered partially with hard working as agricultural laborer, other day laborer employment, and by selling properties or taking loans to do business, cattle farming, or send a household member to temporarily migrate to cities to earn some money. Also, more than 80% of the households in SHS2 could not recover at all, which is highest among the SHS. Looking at socio-economic classes, one can see that more than 70% of poor and moderately poor households could not recover at all (Figure 4.8b). A statistical analysis was performed and it showed that there is no significant difference in recovery from loss by the respondents between the SHS but there exists a significant difference in recovery of loss by socio-economic groups (with $\alpha = 0.05$).

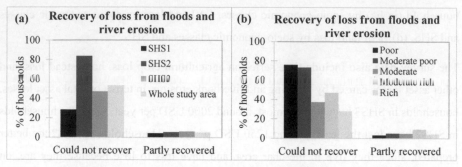

Figure 4.8: Recovery of loss from floods and riverbank erosion in the study area. (a) Recovery of loss from floods and river erosion; (b) Recovery of loss from floods and river erosion.

These statistics do not tell the whole story, since they present averages. During the field survey a few exceptions were found to the overall picture sketched above. While most of the people suffer, a few rich households benefited from flooding and riverbank erosion by giving loans with very high interest to the flood victims. Statistics do not show the richness of individual experiences either, where each household has its own story to tell such as the one presented in Box 4.1. The situation of Mr Abdul Baki Khan is typical for some char dwellers, but the number of times his family moved is also quite exceptional compared to other households.

Box 4.1: An example of a life course of impoverishment in the study area.

Story of Mr. Abdul Baki Khan, an example of living with floods in study area

Mr. Khan, who is 63 years of age, lives in a char of the Jamuna River in Uria union of Fulchari upazila of Gaibandha district. He is a landless farmer whose household consists of seven members. They organize their livelihoods around the river. They grow maize, jute and rice on a field by the Jamuna River char. The occurrence of regular flooding events, associated with eroding land, forces them to relocate every two to three years. Over the period 1978–2018, the family moved 13 times. Because of growing population in the area, finding possible places to live and farm has become increasingly difficult. In the past, he used to have access to some 9 hectares of land (large famer according to the agricultural land size), but today he has become landless. Now he is working on the lands of others and his eldest son is also working with him. His income from agricultural land is often not enough for the family, so they limit their daily needs. From morning to afternoon, he and his son go to the field and work hard. Despite these efforts, their monthly income is only about 50 USD. Their house is made of locally available materials with earthen floor, wood, paddy straw with tin on the top of the house. Mr Khan is an example of many people living in the Jamuna floodplain and its char. Many people like him used to be richer earlier, but they have become poorer because of flooding and erosion.

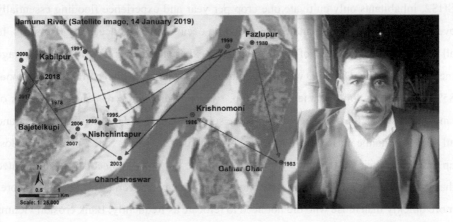

All relocations of Mr. Khan and his family mapped onto current Jamuna River pattern

4.5 DISCUSSION AND CONCLUSIONS

Primary and secondary data were analyzed to reveal how the three SHS and the socio-economic classes are affected by flooding and bank erosion, and their inhabitants adopt adjustment measures to face these hazards and reduce or mitigate the related risks. This study quantified the costs of living with floods and showed that this can lead to general impoverishment in the long term.

While the country (Bangladesh as a whole) has become more prosperous (Khandker, 2007), the per capita gross domestic product (GDP) of Bangladesh is accelerating rapidly (World Bank, 2019a), and the percentage of the population living below the national poverty line is decreasing (World Bank, 2019b). Yet, the prosperity of the study area is below the national level, partly because growing GDP in Bangladesh is mainly urban (BBS, 2018a). According to the National Accounts Statistics for the year 2016 (BBS, 2018b), the per capita annual income of the study area is estimated at 1160 USD compared with the national average of 1544 USD. Based on the surveyed data, it is estimated that per capita annual income in SHS1 at USD 1200, in SHS2 at USD 700, and in SHS3 at USD 1000 (below BBS estimates). It is also estimated that the percentages of households below poverty line is about 18% in SHS1, 32% in SHS2 and 15% in SHS3. Given that the main source of income in this area is represented by farming (Ferdous et al., 2018), the occurrence of flooding in the area highly impacts income generation. In SHS2, inhabitants only cultivate one crop per year and experience flooding essentially every single year. In SHS1 and SHS3, flooding occurs roughly once every two years. In all three SHSs, over 90% of the respondents experienced over the years an average income loss of about 80 USD which they attribute to flooding, which corresponds to about 85% of their monthly income. In addition, riverbank erosion is responsible for loss of land and assets in the area. According to the surveys, households from all three SHS and all socio-economic classes have lost land over the past 50 years. SHS2 people have lost up to 80% of their land and many households in the poorest socio-economic class stated that they have lost all their land. Inhabitants that relocated several times became poorer more quickly than those who did not need to relocate as frequently. Bank erosion victims tend to relocate to nearby land, hoping that they can recuperate their land in the near future. There are no many alternatives to recover from income, land, and asset loss. They either change their occupation from farmer to day laborer, fisherman, or temporarily migrate to cities to earn. However, recovery rate is very low: income loss and asset loss recovery is marginal even for rich class people. Furthermore, due to the lack of employment alternatives in the area, many people need to take loans with high interest during the flood season. In an attempt to increase their loss recovery capacity, people try to save some money for the flood season; however, their savings evaporate rapidly during the flood season when monthly expenses can be higher than incomes. As a result, they move towards impoverishment over time. Household respondents are affected by flooding and riverbank erosion in terms of income losses (Table 4.3), increases in expenses (Table 4.4) as well as land losses (Figure 4.7). The household respondents have also mentioned that it is almost impossible to fully recover the loss (Figure 4.8). This study analysis showed that although flood mortality rates in Bangladesh have been significantly decreasing over time (Mechler and Bouwer, 2015; Kreibich et al., 2017), the costs of adjusting to flooding and riverine erosion are very high and can negatively impact the livelihood of local people. While households mainly blame flooding and bank erosion, there can be other socio-economic factors, such as a lack of investments in alternative types of employment, which also can significantly contribute to the impoverishment of the area.

Figure 4.9 depicts the definitions of coping and adapting introduced in Section 4.1. This research shows that households in the study area are coping (but not adapting) with flooding and riverbank erosion, since the net result of their adjustments is a decline in socio-economic conditions (see orange trajectory in Figure 4.9).

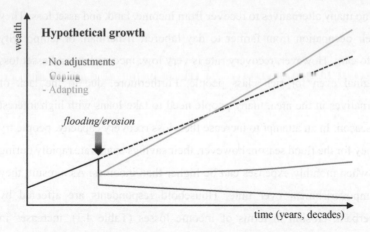

Figure 4.9: Coping or adapting? Deviation from a hypothetical trajectory of socio-economic growth (black line) caused by flooding and erosion, in case of: (i) no adjustments (red line), (ii) coping (orange line), and (iii) adapting (green line).

Loss of homesteads forces people to move to new places without any option and puts them in desperate situations. Despite these extreme living conditions, the char inhabitants do not leave this area because of various obstacles, the lack of available land elsewhere being foremost. They try to stay as long as possible in their home during high floods and in the case of erosion, dismantle it and transport it to another char which is less or not affected by erosion at that moment. While the situation in the char area (SHS2) is extreme, in every location of the study area people are losing income and assets, since they can only partially recover from flood events. Overall, they are getting poorer and therefore less able to make the further adjustments for the next flooding event. Although there are most likely other factors at play, repeated flooding and associated bank erosion contribute to overall impoverishment (in relative terms compared with national figures of economic growth) of the floodplain people of Bangladesh, despite their use of multiple strategies to respond to flooding and riverbank erosion. Due to the scale of works, the flood control measures constructed by the Government are not likely to ever be sufficient to prevent flooding or riverbank erosion. This research posits that indigenous adjustments, such as the adoption of different types of crops to varied flood depths, should be reinforced to reduce the negative impacts of flooding on agriculture and therefore slow down the rate of impoverishment in the area.

This research work does not only contribute to advance the knowledge about socio-hydrological dynamics in Bangladesh, but also provides more general insights for flood risk management in low-income regions of the world. There is an ongoing discussion about the need for a shift from hard (fighting floods) to soft (living with floods) approaches (Opperman et al., 2009). It has been argued that low-income countries such as Bangladesh should not implement hard engineering work to increase levels of structural flood protection, as it has been done in most Western countries, but stick to their traditional softer approach of living with floods (Cook and Lane, 2010). Indeed, some of the polders that were constructed in 1970s had negative impacts on ecosystems and livelihoods and are currently being revised to re-establish a workable sediment and water balance (Ferdous et al., (under review)). However, the Jamuna case study demonstrates that living with floods has enormous costs and it prevents socio-economic growth in the areas. As such, there are no clear-cut answers to the question of how low-income countries should deal with flooding. This research argues that universal recipes do not exist. Instead, there is a need to find trade-offs between hard and soft options depending on the values given by local communities, experts, practitioners, and governments to environmental, social, and economic benefits and costs of alternative strategies. A better understanding of socio-hydrological dynamics, such as the one provided with the Jamuna case study, can help identify these trade-offs by shading light on both the positive and negative effects of living with floods.

5

THE INTERPLAY BETWEEN STRUCTURAL FLOOD PROTECTION, POPULATION DENSITY AND FLOOD MORTALITY ALONG THE JAMUNA RIVER

This chapter is based on an earlier version of the following paper; minor changes were made since the submission of this thesis:

Ferdous, M.R.; Di Baldassarre, G.; Brandimarte, L.; Wesselink, A. (2019). The interplay between structural flood protection, population density and flood mortality along the Jamuna River, Bangladesh. Manuscript is accepted for publication in the Journal Regional Environmental Change.

Abstract

Levees protect floodplain areas from frequent flooding, but they can paradoxically contribute to more severe flood losses. This is known in the literature as the 'safe-development paradox', which was first discussed by Gilbert White in the 1940s. The construction or reinforcement of levees can attract more assets and people in flood-prone area, thereby increasing the potential flood damage when levees eventually fail. Moreover, structural protection measures can generate a sense of complacency, which can reduce preparedness, thereby increasing flood mortality rates. We explore these phenomena in the Jamuna River floodplain in Bangladesh. This is a unique study area as different levels of flood protection have co-existed alongside each other since the 1960s, with a levee being constructed only on the right bank and its maintenance being assured only in certain places. Primary and secondary data on population density, human settlements and flood fatalities were collected to carry out a comparative analysis of two urban areas and two rural areas with different flood protection levels. Our study shows that flood mortality rates in Bangladesh are lower in the areas with lower protection level. Moreover, we found that the higher the level of flood protection, the higher the increase of population density and the number of assets exposed to flooding. This empirical analysis of the unintended consequences of structural flood protection is relevant for the making of sustainable policies of disaster risk reduction and adaptation to climate change in rapidly changing environments.

5.1 INTRODUCTION

In the year 1964, new levees were built to prevent flooding in the village of Char Jabbar, Bangladesh. The presence of this structural protection measure encouraged more human settlements and numerous people moved into this flood-protected area (Burton et al., 1993). A few years later, in November 1970, a tropical cyclone hit Bangladesh, levees were overtopped, and about 6,000 people were killed by flooding (Islam, 1971). The dramatic history of Char Jabbar shows how the net effect of building levees can result into increasing flood losses and fatalities (White, 1945).

Char Jabbar is not an exceptional case. Over the past decades, numerous scholars have shown that structural flood protection tends to be associated with increasing flood exposure, defined here as the population and assets located in flood hazard-prone areas (Jongman et al., 2015), and flood vulnerability, defined here as the susceptibility of the exposed elements to flooding (Jongman et al., 2015). This tendency is typically described as the 'safe development paradox', 'levee effect', 'residual risk', or 'safety dilemma' and it was shown to potentially offset the intended benefits of structural flood protection (e.g. White, 1945; Tobin, 1995; Kates et al., 2006; Burby, 2006; Montz and Tobin, 2008; Scolobig and De Marchi, 2009; Di Baldassarre et al., 2013a,b).

Several studies have shown that increasing the levels of structural flood protection can attract more settlements and high-value assets in the protected areas (e.g. White, 1945; Kates et al., 2006; Montz and Tobin, 2008; Di Baldassarre et al., 2013a,b), thereby increasing exposure to flooding. Kates et al., (2006), for example, discussed that the catastrophic 2005 flooding of New Orleans (Katrina) showed that while flood defence has reduced the negative consequences associate with more frequent events, it also contributed to build up exposure to more rare events. Other studies have explored how structural flood defence can generate a sense of complacency (Tobin, 1995), which can act to reduce preparedness, thereby increasing social vulnerability to flooding (e.g. Burby, 2006; Scolobig and De Marchi, 2009; Ludy and Kondolf, 2012). For instance, Ludy and Kondolf (2012) looked at Sacramento-San Joaquin Delta where the residual risk for lands protected by a 200-year levee is extremely high, but it is completed ignored (or vastly underestimated) by locals. Literature in this field is vast, as shown in the recent review made by Di Baldassarre et al. (2018), and it goes well beyond river and coastal flooding. Logan et al. (2018), for example, analysed tsunami impacts in Taro, Japan. They observed that structural protection measures can cause a false sense of security and encourage development that is vulnerable in the long-term.

The safe-development paradox should not be seen as a mere one-way causal link, but the result of self-reinforcing (bidirectional) feedbacks (Di Baldassarre et al., 2013a,b): e.g. increasing protection levels enable intense urbanisation that will in turn plausibly require even higher protection standards (Viglione et al., 2014). Thus, it can generate lock-in conditions towards exceptionally high levels of flood protection and extremely urbanised floodplains (Di Baldassarre et al., 2018). Whilst this lock-in condition can work in some socioeconomic contexts, such as in The Netherlands (de Moel et al., 2010), it might become unsustainable or socially unjust in other contexts. The costs and benefits of flood protection measures, as well as potential flood losses, are not always fairly shared across social groups (Burton and Cutter, 2008), as seen for instance in the aftermath of the catastrophic 2005 flooding of New Orleans (Masozera et al., 2007).

The recent literature has not only shown how building or raising levees can lead to very intense occupation (with more people and assets than originally expected) of flood-prone areas behind the levee, but also losses of ecological functions (Opperman et al., 2009). However, the narrative that "we need to building higher levees to cope with flooding" remains pervasive not only for policy and decision makers, but also within the scientific community (e.g. Ward et al., 2017). As a result, numerous structural protection structures, such as levees, groynes, spurs or flood-control reservoirs, are being suggested, planned or built in many areas around the world (Di Baldassarre et al., 2018).

Kreibich et al. (2017) explains the reduction of mortality rates in Bangladesh as a result of different factors, including early warning systems based on better flood forecasting (Gain et al., 2015) along with more spontaneous or informal processes, such as the combination of higher education and flood experience leading to increased awareness and preparedness.

This tendency of decreasing flood losses over time is termed 'adaptation effect' in the literature (Di Baldassarre et al., 2015) and it has been observed by other studies (Jongman et al., 2015; Mechler and Bouwer, 2015; Kreibich et al., 2017) across different socio-hydrological contexts. Yet, the literature has also shown that adaptation effects are less significant when the levels of structural flood protection are very high, due to higher reliance (and trust) on levees or flood-control reservoirs (Mård et al., 2018). As such, one of the questions guiding our research work is: how are flood mortality rates and people settled in flood-prone areas influenced by structural flood protection in Bangladesh?

To address this question, we explore different socio-hydrological spaces (Ferdous et al., 2018) in the Jamuna River floodplain in Bangladesh (Figure 5.1). This is a unique study area as different levels of flood protection have co-existed alongside each other since the 1960s, with a levee (i.e. the Brahmaputra Right Embankment, BRE) being constructed only on the right bank and its maintenance being assured only in certain places (Figure 5.1). This consists of four test sites characterised by different levels of structural flood protection (Figure 5.1). Primary and secondary data on population density, human settlements and flood fatalities were collected to carry out a comparative analysis of two urban areas and two rural areas with different flood protection levels.

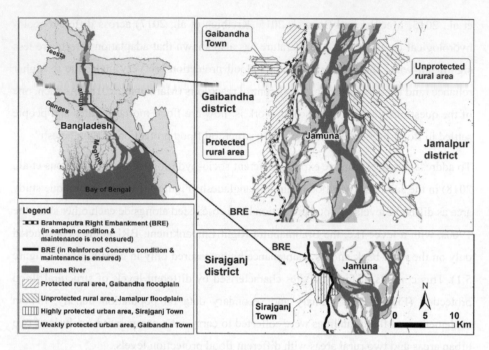

Figure 5.1: Bangladesh and its major rivers. The two insets shows the Brahmaputra Right Embankment (BRE) and the four study areas: the protected rural area in the Gaibandha district, the unprotected rural area in the Jamalpur district, and two urban areas (Gaibandha Town and Sirajganj Town) with different flood protection levee.

5.2 STUDY AREA

In the year 2017, major flooding hit Bangladesh. Almost half (42%) of the country is under water (FFWC/BWDB, 2018), and in numerous island villages along the Jamuna River, entire homes have been washed away, while crops and food supplies all but wiped out. "Villagers described the rains as the worst in living memory" (CNN, 2017). According to FFWC/BWDB (2018), the 2017 flood hit the country twice: on 1st week of July and on 2nd week of August due to excessive rainfall in the upstream of Bangladesh. In both cases, flood duration was about 2 weeks, but the second flood peak in August was more severe. The water level of the Jamuna River crossed the danger level on around 3rd week of August and remained above it for about one month. In the previous 100 years, the highest water level of Jamuna River was the one recorded 20.62 mPWD at

Bahadurabad station in 1988, but such highest water level was exceeded to 20.84 mPWD at the same station in 2017 thereby setting a new flood peak record (FFWC/BWDB, 2018).

Indeed, data of flood losses (about 0.7 million houses and crops of about 0.6 million hectares land were damaged) and fatalities (recorded total is 147) shows that the negative impacts of the 2017 flooding in Bangladesh were massive. Yet, when compared to the most recent events (Figure 5.2), one can observe that flood mortality rates in Bangladesh have been significantly decreasing over time, as previously observed by Mechler and Bouwer (2015).

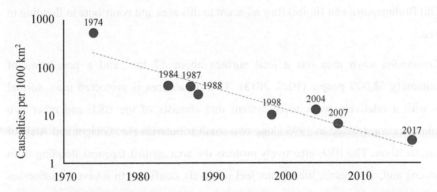

Figure 5.2: Flood fatalities in Bangladesh normalized by flooded area (casualties by 1000 km^2) for major flooding events between 1974 and 2017. (Data source: Brammer (2004); Sultana et al., (2008); Penning-Rowsell et al., (2012); BBS (2016); NDRCC (2017)).

To better understand the interplay between structural protection levels and flood exposure/mortality, we explore the effects of structural flood protection in four different types of human settlements along the Jamuna River floodplain. They consist of two rural environments; the protected rural area in the Gaibandha district (an embankment was constructed in the 1960s parallel to the west bank of the Jamuna River to restrict flood water to enter in that area) and the unprotected rural area (no man-made embankment was constructed along the east bank to restrict flood water) in the Jamalpur district and two urban environments; Gaibandha and Sirajganj with different levels of structural flood protection (Figure 5.1 and Table 5.1).

The protected rural area in the Gaibandha district (right bank of the Jamuna river, Figure 5.1) has a total surface of about 74 km^2 and a population of approximately 111,000 people

105

(BBS, 2013). This rural area is protected by regular annual flooding which is locally termed as normal flooding (Ferdous et al., 2018). However, a few locations of this area are still frequently inundated by excessive rainfall or adjacent small rivers (Alai and Ghagot).

The unprotected rural area in the Jamalpur district (left bank of the Jamuna river, Figure 5.1) has a total surface of about 174 km^2 and a population of approximately 146,000 people (BBS, 2013). As there is no man-made structural protection measure in this rural area, flooding occurs more frequently here than on the right bank. Some other small rivers (e.g. Old Brahmaputra and Jinjira) flow adjacent to this area and contribute to flooding in this area.

The Gaibandha town area has a total surface about 17 km^2 and a population of approximately 68,000 people (BBS, 2013). This urban area is protected from normal floods with a relatively weak levee system that consists of the BRE and other two embankments constructed in 1995 along two small tributaries (i.e. Ghagot and Alai) of the Jamuna River. The BRE effectively protects the area against frequent flooding from the Jamuna and, as a result, inhabitants feel relatively confident to invest in businesses and homesteads (Rahman, 2017, Ferdous et al., 2018). The last extreme flooding events in Gaibandha occurred in 1988 and 2017.

The Sirajganj town area has a total surface about 19 km^2 and a population of approximately 160,000 people (BBS, 2013). This urban area is protected from flooding with a relatively stronger levee system, as the BRE was heightened and reinforced in the 1990s to protect this town from frequent flooding. Still, flooding occurred both in 1988 and 2017. The BRE effectively protects the area against most flooding events from the Jamuna and, as a result, inhabitants feel relatively confident enough to invest in businesses and homesteads (Rahman, 2017).

Table 5.1: Summary of socio-economic factors in the four study areas (BBS, 2013).

	Urban, more protected Sirajganj town	Urban, less protected Gaibandha town	Rural, protected Gaibandha floodplain	Rural, unprotected Jamalpur floodplain
Flood early warning system?	Yes	Yes	Yes	Yes
Disabled population	1.5 %	1.6 %	1.5 %	1.4 %
Sex ratio	1.02	1.00	1.00	0.97
Literacy Rate	63.2 %	74.5 %	31.2 %	30.6 %
Average age	26.7 years	29.5 years	25.0 years	25.2 years
People with electricity facility	90.1 %	84.8%	41.8%	21.0%

5.3 DATA AND METHODS

Our study builds upon previous work about flood risk in Bangladesh (Haque and Zaman, 1993; Brammer, 2010; Cook and Lane, 2010; Cook and Wisner, 2010; Mechler and Bouwer, 2015; Gain et al., 2015; Ferdous et al., 2018). Our analysis is based on secondary data for two urban and two rural areas with different protection levels. In these four study areas, we collected secondary data on population density, satellite images (for human settlements) and flood fatalities and carry out a comparative analysis about the effects of structural flood protection. Time series of the national census from 1974 to 2011 were provided by the Bureau of Statistic of Bangladesh (BBS) and from 1901 to 1961 were provided by Census of Pakistan Population (CPP) (Table 5.2). We used this data to see the increase in population densities in the four study areas. Flood fatalities data were collected from National Disaster Response Coordination Centre (NDRCC) of the Government of Bangladesh (Table 5.2) to analyse the trend of flood fatalities over time. Time series of satellite images of 30m spatial resolution were provided by CEGIS, Bangladesh (Table 5.2). These datasets were used to analyse the expansion of human settlements in the four study areas. Land use and land cover classifications were carried out using optical images with high spectral resolution (7 bands for Landsat 4/5 and 11 bands for Landsat 8). Due to lack of availability of remote sensing data with high spatial resolution prior to 1989, we analysed land use patterns only for the years 1989 and 2014, thereby showing expansion of the human settlement areas over the last three decades. All images were geo-rectified into "Bangladesh Transverse Mercator" (BTM) projection. For better visual interpretation, the false-color composition was used. After visual interpretation, 50 spectral classes were generated using a digital unsupervised classification to derive different land uses and land covers from the satellite images. ERDAS IMAGINE software uses the ISODATA, stands for "Iterative Self-Organizing Data Analysis Technique", algorithm to perform this classification. The ISODATA clustering method uses the minimum spectral distance formula to form clusters. After digital classification, the mixed classes were grouped together, and the similar process was run for refining the classes and increasing accuracy level. The 2014 settlement data were taken from vector data, digitized from multispectral RapidEye (5m of spatial resolution) images. These vector data were converted into raster format with the software ERDAS IMAGINE and used for land use classification.

Table 5.2: Summary of data used for the spatial analysis

Parameter	Years	Information	Source
Satellite images			CEGIS
- Landsat 4 TM	1989	- Spatial resolution 30 m	
		- Spectral resolution 7 bands	
- Landsat 8	2014	- Spatial resolution 30 m	
		- Spectral resolution 11 bands	
Population	1901-2011		CPP (1964); BBS (1974; 1986; 1994; 2005; 2013; 2014a, b)
Flood fatalities	1974-2017		Brammer (2004); Sultana et al., (2008); Penning-Rowsell et al., (2012); BBS (2016); NDRCC (2017)

5.4 RESULTS

Spatial and temporal changes in flood exposure in the two rural areas are depicted in Figure 5.3. In particular, Figure 5.3a shows the spatial distribution of population density in 1961 and 2011, while Figure 5.3b shows density in the period 1961-2011. While both protected and unprotected areas have been increasing since 1961, the results of our study show that protected areas have had a large increase of population density.

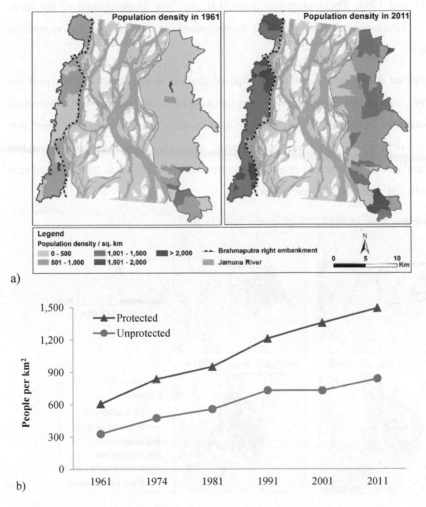

Figure 5.3: Flood exposure in rural areas: a) Population density in 1961 and 2011, b) Population density in the period 1961-2011.

Urban areas are compared in Figure 5.4, which depicts the temporal and spatial evolution of urbanization patterns between 1989 and 2014 (Figure 5.4a) and population density (Figure 5.4b) in the period 1901-2011. Before the construction of the BRE, both urban areas show moderate increase of population, with a similar rate of growth. After the construction of the BRE, both urban areas show a change in the population growth rate: Sirajgani shows a much steeper increase than Gaibandha. The severe effects of major floods occurred in 1987 and 1988 are visible in a drop of population growth between the year 1981 and 1991. These outcomes show that after the reinforcement of the levee system in Sirajgani, the town has had more growth in human population than in Gaibandha.

To corroborate these findings, we also compared changes in population density over the past thirty years in Shahjadpur Upazaila on the west bank of Jamuna river (protected) and in Nagarpur Upazila on the east bank of Jamuna river (unprotected). We found that population density is higher in the protected area (1,730 persons/km^2 vs. 1,100 persons/km^2). Also the increase rate in population density is higher in the protected area (22 persons/km^2/year vs. 10 persons/km^2/year).

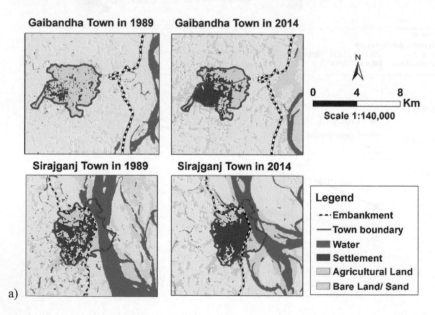

a)

110

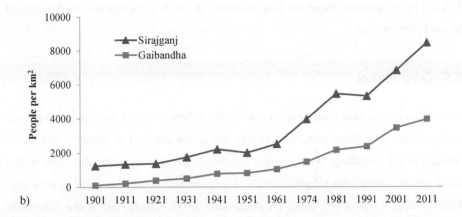

b)

Figure 5.4: Flood exposure in urban areas: a) Land use patterns in 1989 and 2014. b) Population density in the period 1901-2011.

Moreover, the analysis of flood fatalities caused by the 2017 flooding for our study area showed that the flood mortality rates in the areas with lower protection level were less than the flood mortality rates in the areas with higher protection levels: 1 vs. 3 fatalities per 100,000 people in unprotected vs. protected rural areas, and 1 vs. 2 per 100,000 people in less vs. more protected urban areas (Source: National disaster response co-ordination centre (NDRCC), Government of Bangladesh). This difference cannot be explained by different levels of exposure since our spatial analysis, based on the 2017 flood extent map provided by the Dartmouth Flood Observatory (Brakenridge, 2019), showed that the proportions of territory that was flooded in protected and unprotected areas were very similar (59% vs. 55%, respectively).

Secondary data for socio-economic factors are limited for our study area. We select few demographic and socio-economic factors that might influence the population density in our study area. Table 5.1 shows that: i) flood early warning systems are in place in all four test sites, and ii) the sex ratio as well as the proportion of disabled people are relatively homogeneous. This result shows that the prevention of small flooding events via structural measures has not only been associated with more intense urbanisation of flood-prone areas (Figures 5.3 and 5.4), but also with higher mortality rates when extreme flooding events eventually occur. Yet, differences in mortality rates are limited. Moreover, there remain other factors, such as literacy rate, that differ across the four test sites (Table

5.1) and have unknown effects on flood mortality. As such, our empirical results i should be used with caution.

5.5 DISCUSSION

The results of this case study support consolidated theories about the interplay between levels of structural flood protection, people and assets exposed to flooding, and social vulnerability to flooding. While similar outcomes have been broadly discussed in the flood risk literature with reference to US, European and Australian cases studies (e.g. Tobin, 1995; Kates et al., 2006; Di Baldassarre et al., 2018), this is the first study providing empirical evidence of these phenomena in a low-income country. Moreover, the presence of four adjacent study areas with different protection standards enabled an original comparative analysis. As such, the results of this study are relevant for the making of sustainable policies of flood risk reduction and adaptation to climate change in Bangladesh, and inform socio-hydrological models integrating human behaviour in risk analysis (e.g. Sivapalan et al., 2012; Di Baldassarre et al., 2015; Aerts et al., 2018). For instance, we found that more protected areas experience higher flood losses during severe flooding events, but these areas experience less year-by-year damage caused by ordinary floods. Moreover, they have had relatively more economic growth (e.g. access to electricity; Table 5.1), investments, and agricultural incomes (Ferdous et al., 2019). These outcomes can be used to parameterize conceptual models of human-flood interactions (Di Baldassarre et al., 2015) as well as risk assessment methods (Aerts et al., 2018).

Blöschl et al. (2013) distinguish between a top-down rationale for flood risk management, where decisions are based on probabilities of flooding and risk calculation (e.g. cost-benefit analysis), and a bottom-up rationale where the possibility of flooding, social vulnerability and the ability of populations to recover are key for decisions. Our work has unraveled new aspects that can contribute to advance both perspectives in Bangladesh, as the influence of structural flood protection in the historical change of human settlements can improve methods for risk calculation, while the outcomes about mortality rates provide new insights about the link between flood occurrences, preparedness and coping capacities.

Moreover, it is important to note that our findings about the dynamics of human settlements in the Jamuna floodplain are only partly attributable to the combination of the factors presented here, i.e. frequency of flooding events, structural flood protection and household coping capacities. In fact, other external factors, such as migration or lack of alternative settlement locations, may have played an important role in shaping the evolution of the four human settlements analyzed here. As such, more empirical research is needed how endogenous and exogenous factors shape the dynamics of human settlements and contribute to flood risk changes in Bangladesh.

5.6 CONCLUSIONS

A shift from hard (fighting floods) to soft (living with floods) approaches for flood risk management is a general trend in policy and scientific writing today (e.g. Opperman et al., 2009). In terms of policy implications for flood risk management, various scholars have already argued that Bangladesh should not implement hard engineering work and high levels of structural flood protection, but stick to their traditional softer approach (e.g. Haque and Zaman, 1993; Cook and Lane, 2010). In fact, some of the polders that were constructed in 1970s, which had negative impacts on livelihoods and ecosystems, are now being partially removed or revised to re-establish a workable sediment and water balance.

Our work contributes to advance the knowledge underpinning flood risk management in Bangladesh. Yet, there are no clear-cut answers to the question of how should Bangladesh cope with flooding in the coming decades because of the aforementioned complexity of endogenous and exogenous factors. Moreover, the balance between soft and hard approaches also depends on the (unavoidably subjective and different) weights and values given by local people, experts, researchers and governments to economic, environmental and social benefits and costs. There are, in fact, multiple feasible (and desirable) trade-offs between hard and soft approaches and their identification calls for a transparent communication of positive (often intended) and negative (often unintended) effects of alternative measures in flood risk management.

6

CONCLUSIONS

Flood is one of the main hazards in the world. Managing this hazard is a big societal challenge. While working on flooding and riverbank erosion during my professional life, I deeply felt the need for the kinds of scientific knowledge that would help managers and decision makers to better deal with flood management or to design more adequate interventions and improve their decisions. Hence, a main objective of the research presented in this thesis was to contribute to producing such knowledge. The thesis set out to explore the phenomena, opportunities and risks generated by the interactions between physical and societal processes along the Jamuna River in Bangladesh. I conceptualize these interactions as temporally dynamic and spatially diverse combinations of fighting and living with water. In this chapter, I discuss and assess whether or to what extent I met my objectives and reflect on the implications of my findings for socio-hydrological research and flood risk management in Bangladesh.

6.1 KNOWLEDGE ADDED TO SOCIO-HYDROLOGICAL STUDIES AND LIMITATIONS TO THE RESEARCH

The scientific understanding of how and to what extent hydrological processes influence or trigger changes in societal processes (and vice-versa) is not yet very advanced. This is because until recently, the study of riverine processes tended to be carried out in separation from the study of societal processes – with both types of studies relying on distinct theoretical and methodological traditions and approaches. This thesis engaged with an emerging body of so-called socio-hydrological research by explicitly acknowledging the interactions and feedback loops between water and human systems. I conducted my research in Bangladesh, one of the largest and most densely populated deltas in the world. My study contributed to socio-hydrological scholarship. First, through its production of a large dataset that allowed combining and correlating data about flood events with information about people's behavior. It provided a chance to test whether socio-hydrological thinking – usually applied to understand and predict socio-hydrological processes at larger spatial and temporal scales – can be used for more detailed mapping of the interactions between society and rivers in a specific location. I summarize the outcomes of this test in the next section 6.2.

To select the study area and collect the necessary primary data, I made used of, and benefited from, my professional experience of more than 10 years with flooding and river bank erosion studies along the Jamuna floodplain. During this time, I became intimately acquainted with the activities and behaviors of the local people during flooding events. For the present study, I collected survey data of approximately 900 households, interviewing them on their flooding and riverbank erosion experiences and their responses to these events. I collected the location of human settlements and agricultural lands, the main occupation, age, income and expenditures of household members, the wealth and origin of the households, together with information on different flood experiences such as of flood levels, frequency, duration, damages, effects on agricultural income and expenditures, as well as flood adjustment options (including migration) and experiences with river erosion. I collected these data from about a 500 km^2 territory along the Jamuna River, i.e. chars inside the river banks and villages along both (left and right) riparian areas. The richness of my dataset is a main asset of this study. By doing my surveys and listening the life stories of the local people, I observed that the differences in the physical conditions within the study area are associated with different ways of living with or adapting to floods. This observation, coupled with my past experiences and confirmed by the outcomes of the reconnaissance survey, helped me refine socio-hydrology by linking larger patterns in flood-society dynamics (fighting and/or adapting) to direct experiences of flooding.

When I started my PhD research, I intended to explore the socio-hydrological interactions in both ways. In retrospective, I conclude that I have not succeeded with that ambition because of the small size of the study area. Many of the decisions about flood management are taken at a higher, national level. My study does clearly show the impacts of local hydrological processes on local social processes, but it does not reveal how local social processes affect local hydrology. This is not to deny that in some places inhabitants try to influence local hydrology. During the household surveys, the respondents along the west bank for instance mentioned that the spurs that were constructed by the Government about 20 km upstream increased the rate of erosion in their neighborhood. They tried to install locally made erosion control measures, constructed out of bamboo. These attempts

117

to stop erosion failed because they are not significant compared with the natural forces of this large river.

This discussion leads to a more generic question about socio-hydrology: at what scale should (or can) socio-hydrology be studied, and how does this relates with the purpose of the study? If decisions are taken at the national scale, is the national scale the most appropriate for socio-hydrology research? After conducting this research, I do believe that to understand the effects of hydrology on local social processes, a small area like my case study area is appropriate. It does nevertheless require detailed and intensive data collection efforts that should preferably be combined with local knowledge. However, if the aim is to explore the social effects on hydrological processes, there may be merit in choosing a larger study area, e.g. the entire river basin. This is because, from a physical perspective, flooding and the associated erosion originate in upstream areas and their characteristics are due to intensive rainfall events and other human interventions exacerbating or alleviating flood peaks. From a societal perspective, at least in Bangladesh, human interventions to cope with flooding and erosion are of such a scale that they cannot be initiated and implemented locally: they are decided at a national level. To explore the social effects on hydrological processes in Bangladesh, different hypotheses and different data are therefore needed.

6.2 SYNTHESIS OF THE RESEARCH

In my thesis, I introduced a new concept to operationalize the search for patterns of specific cases: socio-hydrological spaces (SHSs) as a useful 'tool' for the study of the socio-hydrology of floodplains. SHSs help make the necessary intermediary step between the messy reality of the specific location (space) and the abstractions of conceptual and mathematical models. With this tool, it becomes possible to create a middle ground where the variability of reality and the unpredictability of human behavior and decisions are preserved and not force-fitted into a model, while at the same time patterns (due to combinations of similar or comparable fight and/or adapt responses) can be recognized. The primary function of SHSs is that it can serve as a lens through which to view and make sense of the complex reality of specific cases in order to relate these to larger patterns in human-river interactions. Such patterns can then be compared and contrasted to patterns in other locations. These patterns can then be generalized through the more formal conceptualization of socio-hydrological systems. On the one hand SHSs thereby relate to a specific space, on the other hand they help to find general patterns of human-river interactions. SHSs entail an invitation to the researcher to have an open mind to the existence of unexpected patterns in location-specific data and to take prior knowledge and experience of the location seriously: society, history, economics, natural system, technical interventions, etc.

I applied the concept of SHSs to a floodplain area along the Jamuna River in Bangladesh, illustrating and showing evidence for the existence of such spaces in the study area. Because of the differential treatment of the right and left bank, and the existence of chars in the middle of the river, three distinct socio-hydrological spaces were identified in the study area; (i) floodplain areas protected with flood embankment; (ii) chars (river islands) within the river banks and (iii) floodplain areas separated by the main river only by natural levees. Each SHS shows distinct features in terms of flood-society interactions, showing that the dynamic interaction of floods and society depends on different combinations of hydrological and societal characteristics along the Jamuna River. Perceptions of flooding, flood frequencies and (histories and memories of) flood damages are all different in these three spaces. Riverbank erosion is experienced in each zone, but mainly by inhabitants in the dynamic char areas.

Socio-hydrological space is a concept that can enrich the study of socio-hydrology because it allows a more detailed understanding of and attention to human-water interactions in a specific location. Such attention is useful anywhere in the world and for other socio-hydrological systems than floodplains. The use of SHSs will also make socio-hydrological analyses more policy-relevant: SHS gives researchers incentives to actually go to the field, talk to inhabitants and officials, and obtain a thorough understanding of the often messy specifics of the location. This in itself is an important addition to the formal equations and models of socio-hydrological analyses. In terms of practical use, SHS can for instance be added as additional element to rapid rural appraisals, or other social assessments, to draw attention to how material conditions (hydrological and

119

technical/infrastructure) co-shape social situations. This would be useful for developing interventions under disaster management, but also other development goals.

A second conceptual contribution of the thesis consists in my testing of the 'levee effect' for the studied area. The 'levee effect' was first noted by White (1945). He asserted that the construction of a levee (or other types of structural measure of flood protection) to protect a certain area from frequent flooding might attract more people and more assets in the area, thereby increasing the potential flood damages when the levee eventually fails. Paradoxically, flood risk may thereby rise as a consequence. I explored the 'levee effect' in three districts along the Jamuna River by comparing the historical socio-economic development and the current state of affairs. I compared areas distinguished by different protection levels (good, moderate and none) and by dominant economic activity (rural or urban). When I was travelling for my field work in the study area, the levee effect was clearly apparent even before doing any statistical analysis. Still I needed and wanted to quantify it, and I decided to make use of my rich empirical data to do this. Overall, the analysis clearly shows how different protection levels co-shaped socio-economic development along the Jamuna River. In the rural areas, the analysis confirmed that the flood losses were often lower in the protected area, but higher during the most severe events. Population density, average household income and wealth, expansion of settlement areas and transport infrastructure are higher in the more protected areas. In the two urban areas with good and moderate protection level, the rates of expansion of settlement areas, population density, investments and transport infrastructures show expected differences. Overall, flood protection measures reduced flood frequency and allowed for more urbanization and faster economic growth in urban areas, and higher incomes and lower losses in rural areas. In rural areas the differences in economic development are statistically significant.

Despite the high residual risk, people feel safe and confident to invest because they feel protected by levees, which in turn speeds up economic development. On the other hand, residents in the unprotected area do not feel safe against flooding. During my field work, I did not observe as much private investment as in the protected areas. Many people accepted flooding as a fate of life. They accept the risk because they have no other place to go, and expect to recover their losses by continuing their business after a flood. The analysis showed that while levees can contribute to the economic development of Bangladesh, the construction and maintenance of levees along the Jamuna River is very costly. As such, levee breaches are not so rare while river erosion continue to be very frequent.

A third area of investigation concerned the understanding of how people are living with flooding along the Jamuna floodplain in Bangladesh and how this affects their livelihoods. Bangladeshi people have always lived with floods. They made adjustments to their homesteads and livelihoods to cope as best as possible with floods and riverbank erosion. The residents of the study area are affected by flooding and river bank erosion, and all adopt almost similar measures to face these hazards and reduce or mitigate the related risks. However, my analysis shows that the mere fact that people adjust their livelihoods to natural hazards should not lead to the false conclusion that they can adapt and thrive in the long term. By quantifying the costs of living with floods, I showed that it leads to general impoverishment in the long run. While the country Bangladesh as a whole has become more prosperous, the prosperity of the study area is below the national level. The main source of income in the study area is agricultural farming. The occurrence of flooding and river bank erosion in the area highly impacts the possibility to generate farm income. Many residents who used to be rich when they were younger, became gradually poorer because of consecutive flooding and river bank erosion. Land loss due to river bank erosion is the main causal factor of this. Land erosion makes people relocate. Inhabitants who had to relocate several times became poorer more quickly than those who did not need to relocate as frequently. Residents are facing income loss followed by crop and land losses due to flooding and riverbank erosion. At the same time, yearly expenses are increasing. Their savings evaporate rapidly during the flood season when monthly expenses are higher than incomes. There are not many alternatives to recover from income, land and asset loss and the recovery rate is very low even for the rich people. As a result, they impoverish over time.

Although flooding and bank erosion are the two major sources of impoverishment in the area, socio-economic factors also contribute to this development, such as a lack of investments in alternative employment. Furthermore, the statistics do not tell the whole story, since they present averages. I conclude that although the Government constructed

121

flood control measures to prevent flooding or riverbank erosion, these interventions are not likely to be sufficient to prevent further impoverishment in the long term, let alone to promote the development of the study area.

A last theme of my PhD research concerned the interplay of flood vulnerability and structural protection levels in three district along the Jamuna River. The construction of levees can attract more assets and people in flood-prone area, thereby increasing the potential flood damage when levees fail or breach. Moreover, due to structural protection measures people feel safe which can reduce preparedness, thereby increasing flood mortality rates. Through a comparative analysis of four adjacent areas with different protection levels, I explored these phenomena. The results support consolidated theories about the interplay between levels of structural flood protection, people and assets exposed to flooding, and social vulnerability to flooding. More protected areas experience higher flood losses during extreme flooding events, these areas experience less year-by-year damage caused by ordinary floods. Mortality rates due to floods are higher in the protected area than in unprotected area.

In flood risk management, the current approach is to move from fighting with floods to living with floods. Various scholars have suggested that Bangladesh should stick to the traditional approach of living with floods, proposing that Bangladesh needs to better use its indigenous knowledge. Yet, there are no clear-cut answers to the question of how Bangladesh should deal or cope with flooding in the coming decades. The balance between soft and hard approaches also depends on the values given by local people, experts, researchers and governments to economic, environmental and social benefits and costs.

6.3 POLICY RELEVANCE FOR FLOOD RISK MANAGEMENT

Flood events during the monsoon season are the norm in Bangladesh, as most of the country resides in the deltaic floodplains of three of the biggest rivers of the world: the Ganges, the Brahmaputra and the Meghna. The duration and extent of flooding varies each year according to maximum water levels in these rivers and the intensity of rainfall upstream of, and within, Bangladesh (Brammer, 2004). After devastating floods in the

1950s, successive governments took several initiatives to control and minimize flood damages. At first, they setup a Flood Commission in 1955 and a UN technical assistance team (the Krug mission) in 1956, to examine possible flood control measures. Later on, in 1964, the then government formulated a Master Plan to provide flood protection to agricultural lands (IECo, 1964). Most of the planned flood control and irrigation projects were implemented in the 1970s and 1980s (Brammer, 2004). The Flood Action Plan (FAP) was initiated in 1990s with the aim of a better flood management by means of large engineering interventions (Sultana et al., 2008). Since 1950s the policies for flood management have inclined towards fighting floods. Bangladesh has formulated several policies for flood risk management with the help of donor support and also gained wide experience in flood risk management. In spite of all these efforts, contemporary flood risk management strategies of Bangladesh are still not up to the mark as floods continue to cause lots of sufferings, deaths, migration and asset loss to the local people (Chowdhury, 2000; BDER, 2004; CPD, 2004; Best, Ashworth, Sarker & Roden, 2007; IOM, 2010). After and through conducting this thesis, and also based on my working experience, I have come to the conclusion that one important cause of this lack of effectiveness of past measures is the gap between the floodplain people and policymakers: there is very little sharing of knowledge, collaboration in solving problems or implementing flood management policies. This is partly because of a distinct preference of the Government for hard engineering solutions, those that require advanced expertise, over so-called softer measures which require the patience to listen to and learn from what those who experience floods are doing and thinking. I hope that my thesis helps underscore the importance of local, indigenous knowledge: the importance of taking the many years of experiences of the local people of living with and adapting to floods and of understanding the river seriously when developing flood management interventions. Effective listening to people will improve understanding of the interactions between hydrological and social processes along the floodplains, which in turn will create a better basis for developing effective policies. I have learned from my thesis that socio-hydrological study can minimize the gap and help a better flood risk management as it deals with long-term historical hydrological data and as well as long-term social processes.

The Government of Bangladesh can develop policy from two perspectives: the national one, focusing on national economic development, and the local one, i.e. focusing on the wellbeing and livelihoods of rural and poor populations as it differs for different locations. Ideally, these policies should complement each other. Blöschl et al. (2013) provide the inspiration for this distinction. They distinguish between a top-down or economic rationale for flood risk management where decisions are based on probabilities of flooding and (economic) risk calculation (and/or cost-benefit), and a bottom-up or social rationale where the possibility of flooding and the ability of populations to recover are key for decisions. The latter may result in strategies to recognize and support the coping strategies that I observed in my research or may help identify ways to prevent that floods provoke gradual impoverishment. The former normally produces technical interventions, such as the massive levees that have proven not to be technically sustainable and that ignore social realities. A better understanding of socio-hydrological dynamics, such as the one provided with my case study, can help identify these trade-offs by shedding light on both the positive and negative effects of living with floods. I observed both of these effects in my study area. In sum, there is no clear-cut answer as to how far the government should opt for hard engineering or soft measures: both can and will be needed. What is requires is a better appreciation of how 'hard' and 'soft' dynamically interact in producing outcomes, so that interventions can become more effective.

REFERENCES

Adger, W.N.: Vulnerability. Global Environmental Change, 16, 268–281, 2006.

Aerts, J. C. J. H., Botzen, W. J., Clarke, K. C., Cutter, S. L., Hall, J. W., Merz, B., Michel-Kerjan, E., Mysiak, J., Surminski, S. and Kunreuther, H.: Integrating human behaviour dynamics into flood disaster risk assessment, Nature Climate Change, 8, 193-199, doi: 10.1038/s41558-018-0085-1, 2018.

Alamgir, M.: Famine in South Asia. 1st ed.; Oelgeschlager, Gunn & Hain, Cambridge, UK, 1980.

Ali, A.: Climate Change Impacts and Adaptation Assessment in Bangladesh, Climate Research, 12, 109-116, doi:10.3354/cr012109, 1999.

Allison, M. A.: Historical Changes in the Ganges-Brahmaputra Delta Front, J. Coast. Res., 14, 1269–1275, 1998.

Ayeb-Karlsson, S., van der Geest, K., Ahmed, I., Huq, S., Warner, K.: A people-centred perspective on climate change, environmental stress, and livelihood resilience in Bangladesh, Sustainability Science, 11, 679-694, 2016.

Banerjee, L.: Creative destruction: Analysing flood and flood control in Bangladesh, Environmental Hazards, 9, 102-117, 2011.

Barendrecht, M. H., Viglione, A., and Blöschl, G.: A dynamic framework for flood risk, Water Security, 1, 3–11, https://doi.org/10.1016/j.wasec.2017.02.001, 2017.

Barnett, K., Mercer, S. W., Norbury, M., Watt, G., Wyke, S., and Guthrie, B.: Epidemiology of multimorbidity and implications for health care, research, and medical education: a cross-sectional study, The Lancet, 380, 37-43, 2012.

Barnett, J.: O'Neill, S. Maladaptation. Glob. Environ. Chang. 2010, 2, 211–213.

BBS: Bangladesh Population Census 1974, Village Population Statistics, Bangladesh Bureau of Statistics, Dhaka, 1974.

BBS: Bangladesh Population Census 1981, Community Tables of all Thanas, Bangladesh Bureau of Statistics, Dhaka, 1986.

BBS: Bangladesh Population Census 1991, Socio-economic and demographic report, Bangladesh Bureau of Statistics, Dhaka, 1994.

BBS: Bangladesh Population and Housing Census 2001, Community report, Bangladesh Bureau of Statistics, Dhaka, 2005.

BBS: Bangladesh Population & Housing Census-2011, Community Report: Gaibandha, Bangladesh Bureau of Statistics, Dhaka, Bangladesh, 1-547, 2013.

BBS: Bangladesh Population & Housing Census-2011, Community Report: Jamalpur, Bangladesh Bureau of Statistics, Dhaka, Bangladesh, 1-654, 2014a.

BBS: Bangladesh Population & Housing Census-2011, Community Report: Sirajganj, Bangladesh Bureau of Statistics, Dhaka, Bangladesh, 1-946, 2014b.

BBS: Economic Census 2013, District Report Gaibandha, Bangladesh Bureau of Statistics, Dhaka, Bangladesh, 1-472, 2016a.

BBS: Economic Census 2013, District Report Jamalpur, Bangladesh Bureau of Statistics, Dhaka, Bangladesh, 1-518, 2016b.

BBS: Bangladesh disaster related statistics 2015, Climate change and natural disaster perspectives, Bangladesh Bureau of Statistics (BBS), Dhaka, 2016c.

BBS: Statistical year book 2016, Bangladesh Bureau of Statistics, Dhaka, Bangladesh, 2016d.

BBS: Yearbook of Agricultural Statistics-2017, 29th ed.; Bangladesh Bureau of Statistics: Dhaka, Bangladesh, 2018a; pp. 1-576.

BBS: National Accounts Statistics (Provisional Estimates of GDP, 2017-18 and Final Estimates of GDP, 2016-17), Bangladesh Bureau of Statistics: Dhaka, Bangladesh, 1-88, 2018b.

Bangladesh Disaster and Emergency Sub-Group (BDER): Monsoon Floods 2004: Post Flood Needs Assessment Summary Report, 36 pp, Dhaka, Bangladesh (see http://www.reliefweb.int/w/rwb.nsf/0/0601496727bb568ac1256f230033fbc5 and http://www.reliefweb.int/library/documents/2004/lcgbang-6oct.pdf), 2004.

Berman, R., Quinn, C., and Paavola, J.: The role of institutions in the transformation of coping capacity to sustainable adaptive capacity, Environmental Development, 2, 86–100, 2012.

Best, J. L., Ashworth, P. J., Sarker, M. H., and Roden, J. E.: The Brahmaputra-Jamuna River, Bangladesh, Large rivers: geomorphology and management, 395-430, 2007.

Black, R., Kniveton, D., & Schmidt-Verkerk, K: Migration and climate change: Toward an integrated assessment of sensitivity. In Disentangling Migration and Climate Change (pp. 29-53), 2013.

Blair, P. and Buytaert, W.: Socio-hydrological modelling: a review asking "why, what and how?", Hydrol. Earth Syst. Sci., 20, 443–478, https://doi.org/10.5194/hess-20-443-2016, 2016.

Blöschl, G., Nester, T., Komma, J., Parajka, J. and Perdigão, R. A. P.: The June 2013 flood in the Upper Danube Basin, and comparisons with the 2002, 1954 and 1899 floods, Hydrology and Earth System Sciences, 17, 5197-5212, doi: 10.5194/hess-17-5197-2013, 2013.

Brakenridge G.R.: "DFO Flood Event 4459", Dartmouth Flood Observatory, University of Colorado, Boulder, Colorado, USA, https://floodobservatory.colorado.edu/Events/2017Bangladesh4459/2017Bangladesh4459.html. August 2019.

Brammer, H.: Floods in Bangladesh: II. Flood mitigation and environmental aspects, The Geographical Journal, 156, 158-165, 1990.

Brammer, H.: Can Bangladesh be Protected from floods?, The University Press Ltd., Dhaka, 1–262, 2004.

Brammer, H.: After the Bangladesh Flood Action Plan: Looking to the future, Environmental Hazards, 9, 118-130, doi: 10.3763/ehaz.2010.SI01, 2010.

Brouwer, R., Akter, S., Brander, L. and Haque, E.: Socioeconomic vulnerability and adaptation to environmental risk: a case study of climate change and flooding in Bangladesh, Risk Analysis An International Journal, 27, 313-326, 2007.

Burby, R. J.: Hurricane Katrina and the Paradoxes of Government Disaster Policy: Bringing About Wise Governmental Decisions for Hazardous Areas, The ANNALS of the American Academy of Political and Social Science, 604, 171-191, doi: 10.1177/0002716205284676, 2006.

Burton, C. and Cutter, S. L.: Levee failures and social vulnerability in the Sacramento-San Joaquin Delta area, California, Natural Hazards Review, 9(3), 136-149, 2008.

Burton, I., Kates, R. W., and White, G. F.: The human ecology of extreme geophysical events. FMHI Publications, Paper 78, 1-33, 1968.

Burton, I., Kates, R. W., White, G. F.: The environment as hazard, The Guilford Press, New York/London, 1993.

BWDB: Environmental Assessment/Analysis Reports: River Training Studies of the Brahmaputra River (Report E 0001), Dhaka: Bangladesh Water Development Board, 1992.

Castree, N., Adams, W. M., Barry, J., Brockington, D., Büscher, B., Corbera, E., Demeritt, D., Duffy, R., Felt, U., Neves, K., Newell, P., Pellizzoni, L., Rigby, K., Robbins, P., Robin, L., Rose, D. B., Ross, A., Schlosberg, D., Sörlin, S., West, P., Whitehead, M., and Wynne, B.: Changing the intellectual climate, Nat. Clim. Change, 4, 763–768, https://doi.org/10.1038/nclimate2339, 2014.

CEGIS: Long-term Erosion Process of the Jamuna River, Jamuna- Meghna River Erosion Mitigation Project, Bangladesh Water Development Board, Dhaka, 1–73, 2007.

CEGIS: Prediction of River Bank Erosion along the Jamuna, the Ganges, the Padma and the Lower Meghna Rivers in 2014, Bangladesh Water Development Board, Dhaka, 1-76, 2014.

CEGIS: Prediction of River Bank Erosion along the Jamuna, the Ganges and the Padma Rivers in 2016, Bangladesh Water Development Board, Dhaka, 1–72, 2016.

Ceola, S., Laio, F., and Montanari, A.: Satellite nighttime lights reveal increasing human exposure to floods worldwide, Geophysical Research Letters, 41, 7184-7190, 2014.

Chen, X., Wang, D., Tian, F., and Sivapalan, M.: From channelization to restoration: Sociohydrologic modeling with changing community preferences in the Kissimmee River Basin, Florida, Water Resour. Res., 52, 1227–1244, https://doi.org/10.1002/2015WR018194, 2016.

Chowdhury, M.: The 1987 flood in Bangladesh: an estimate of damage in twelve villages, Disasters, 12, 294-300, 1988.

Chowdhury, M.R.: An assessment of flood forecasting in Bangladesh: the experience of the 1998 flood, Natural Hazards, 22, 139–169, 2000.

Ciullo, A., Viglione, A., Castellarin, A., Crisci, M., and Di Baldassarre, G.: Socio-hydrological modelling of floodrisk dynamics: comparing the resilience of green and technological systems, Hydrol. Sci. J., 62, 880–891, https://doi.org/10.1080/02626667.2016.1273527, 2017.

CNN: A third of Bangladesh under water as flood devastation widens. Updated September 1, 2017, https://edition.cnn.com/2017/09/01/asia/bangladesh-south-asia-floods/index.html (Accessed 07 November 2018), 2017.

Coleman, J.M.: Brahmaputra River: channel processes and sedimentation, Sediment. Geol., 3, 129-239, doi:org/10.1016/0037-0738(69)90010-4, 1969.

Collenteur, R. A., de Moel, H., Jongman, B., and Di Baldassarre, G.: The failed-levee effect: Do societies learn from flood disasters?, Natural Hazards, 76, 373-388, 2015.

Cook, B. R. and Wisner, B.: Water, risk and vulnerability in Bangladesh: Twenty years since the FAP, Environmental Hazards, 9, 3-7, doi:10.3763/ehaz.2010.SI09, 2010

Cook, B. R. and Lane, S. N.: Communities of knowledge: Science and flood management in Bangladesh, Environmental Hazards, 9, 8-25, doi: 10.3763/ehaz.2010.SI06, 2010.

Centre for Policy Dialogue (CPD): Rapid Assessment of Flood 2004: Interim Report, August 12 2004, Dhaka, 47 pp. (see http://www.cpd-bangladesh.org/flood.pdf and http://www.cpd-bangladesh.org), 2004.

CPP: Census of Pakistan Population 1961: Volume 2, East Pakistan, Tables & Report, Ministry of Home & Kashmir Affairs, Karachi, 1964.

de Moel, H., Aerts, J. C. J. H. and Koomen, E.: Development of flood exposure in the Netherlands during the 20th and 21st century, Global Environmental Change, 21, 620-627, doi: 10.1016/j.gloenvcha.2010.12.005, 2010.

Di Baldassarre, G., Castellarin, A., Montanari, A., and Brath, A.: Probability-weighted hazard maps for comparing different flood risk management strategies: a case study. Natural Hazards, 50, 479-496, 2009.

Di Baldassarre, G., Montanari, A., Lins, H., Koutsoyiannis, D., Brandimarte, L., and Bölschl, G.: Flood fatalities in Africa: From diagnosis to mitigation, Geophys. Res. Lett., 37, 2–6. https://doi.org/10.1029/2010GL045467, 2010.

Di Baldassarre, G., Kooy, M., Kemerink, J. S., and Brandimarte, L.: Towards understanding the dynamic behaviour of floodplains as human-water systems, Hydrol. Earth Syst. Sci., 17, 3235–3244, https://doi.org/10.5194/hess-17-3235-2013, 2013a.

Di Baldassarre, G., Viglione, A., Carr, G., Kuil, L., Salinas, J. L., and Blöschl, G.: Socio-hydrology: conceptualising human flood interactions, Hydrol. Earth Syst. Sci., 17, 3295–3303, https://doi.org/10.5194/hess-17-3295-2013, 2013b.

Di Baldassarre, G., Kemerink, J. S., Kooy, M., and Brandimarte, L.: Floods and societies: the spatial distribution of water-related disaster risk and its dynamics, Wiley Interdisciplinary Reviews: Water, 1(April), 133–139, https://doi.org/10.1002/wat2.1015, 2014.

Di Baldassarre, G., Viglione, A., Carr, G., Kuil, L., Yan, K., Brandimarte, L., and Blöschl, G.: Debates – Perspectives on socio-hydrology: Capturing feedbacks between physical and social processes, Water Resour. Res., 51, 4770–4781, https://doi.org/10.1002/2014WR016416, 2015.

Di Baldassarre, G., Saccà, S., Aronica, G. T., Grimaldi, S., Ciullo, A., and Crisci, M.: Human-flood interactions in Rome over the past 150 years, Adv. Geosci., 44, 9–13, https://doi.org/10.5194/adgeo-44-9-2017, 2017.

Di Baldassarre, G., Kreibich, H., Vorogushyn, S., Aerts, J., Arnbjerg-Nielsen, K., Barendrecht, M., Bates, P., Borga, M., Botzen, W., Bubeck, P., De Marchi, B., Llasat, C., Mazzoleni, M., Molinari, D., Mondino, E., Mård, J., Petrucci, O., Scolobig, A., Viglione, A., Ward, P.J.: Hess Opinions: An interdisciplinary research agenda to explore the unintended consequences of structural flood protection, Hydrology and Earth System Sciences, 22, 5629–5637, 2018.

Domeneghetti, A., Carisi, F., Castellarin, A., and Brath, A.: Evolution of flood risk over large areas: Quantitative assessment for the Po river, Journal of Hydrology, 527, 809-823, 2015.

Eakin, H., and Appendini, K.: Livelihood change, farming, and managing flood risk in the Lerma Valley, Mexico, Agriculture and Human Values, 25, 555-566, 2008.

Elahi, K.M. (1972), Urbanization in Bangladesh: A Geodemographic Study. Oriental Geographer 16(1): 39–48.

Elshafei, Y., Sivapalan, M., Tonts, M., and Hipsey, M. R.: A prototype framework for models of socio-hydrology: identification of key feedback loops and parameterisation approach, Hydrol. Earth Syst. Sci., 18, 2141–2166, https://doi.org/10.5194/hess-18-2141-2014, 2014.

FAP 16/19: Charland Study Overview: Summary Report. (Prepared by Irrigation Support Project for Asia and the Near East - Flood Action Plan components 16&19), Flood Plan Coordination Organisation, Ministry of Irrigation, Dhaka. 1993.

Fenton, A., Paavola, J. and Tallontire, A.: Autonomous adaptation to riverine flooding in Satkhira District, Bangladesh: implications for adaptation planning, Regional Environmental Change, 17, 2387-2396, 2017.

Ferdous, M. R.: Floods and Society: going beyond borders: Understanding the dynamics of the southwest coastal region (Satkhira-Khulna-Bagerhat) of Bangladesh as a coupled human-water system, MSc Thesis UNESCO-IHE, WSE-HWR-14.11 by Md Ruknul Ferdous, Delft, The Netherlands, 2014.

Ferdous, M. R., Wesselink, A., Brandimarte, L., Slager, K., Zwarteveen, M., & Di. Baldassarre, G.: Socio-hydrological spaces in the Jamuna River floodplain in Bangladesh. Hydrology and Earth System Sciences, 22, 5159-5173, 2018.

Ferdous, M. R.: Data set to: Socio-hydrological Spaces in the Jamuna River floodplain in Bangladesh, https://doi.org/10.17605/OSF.IO/C7RQY, 2018.

Ferdous, M.R.: Wesselink, A.; Brandimarte, L.; Slager, K.; Zwarteveen, M. and Di Baldassarre, G. (2019). The Costs of Living with Floods in the Jamuna Floodplain in Bangladesh. Water, 11, 1238, https://doi.org/10.3390/w11061238.

Ferdous, M. R.: Wesselink, A.; Brandimarte, L.; Di Baldassarre, G. and Rahman, M. M. (2019). The levee effect along the Jamuna River in Bangladesh. Water International, https://doi.org/10.1080/02508060.2019.1619048.

Ferdous, M.R.: Di Baldassarre, G.; Brandimarte, L.; Wesselink, A. Exploring the interplay of flood vulnerability and structural protection levels in Bangladesh. Regional Environmental Change (under review).

FFWC/BWDB: Annual Flood Report 2016, Flood Forecasting and Warning Centre, Bangladesh Water Development Board, Dhaka, 1-88, 2017.

FFWC/BWDB: Annual Flood Report 2017, Flood Forecasting and Warning Centre, Bangladesh Water Development Board, Dhaka, 1-108, 2018.

Findlay, A. M.: Migration: flooding and the scale of migration, Nature Climate Change, 2, 401-402, 2012.

Fox-Rogers, L., Devitt, C., O'Neill, E., Brereton, F., and Clinch, J. P.: Is there really "nothing you can do"? Pathways to enhanced flood-risk preparedness, Journal of Hydrology, 543, 330-343, 2016.

Gain, A. K., Mojtahed, V., Biscaro, C., Balbi, S. and Giupponi, C.: An integrated approach of flood risk assessment in the eastern part of Dhaka City, Nat. Hazards, 29, 1499–1530, doi: 10.1007/s11069-015-1911-7, 2015.

Gallopin, G. C.: Linkages between vulnerability, resilience, and adaptive capacity, Global Environmental Change, 16, 293–303, 2006.

Gober, P. and Wheater, H. S.: Socio-hydrology and the science-policy interface: a case study of the Saskatchewan River basin, Hydrol. Earth Syst. Sci., 18, 1413–1422, https://doi.org/10.5194/hess-18-1413-2014, 2014.

Goodbred, S.L. Jr. and Kuehl, S. A.: Holocene and Modern Sediment Budgets for the Ganges-Brahmaputra River System: Evidence for Highstand Dispersal of Floodplain, Shelf, and Deep-Sea Depocenters, Geology, 27, 1999.

Goodbred, S. L., Kuehl, S. A., Steckler, M. S., and Sarker, M. H.: Controls on facies distribution and stratigraphic preservation in the Ganges – Brahmaputra delta sequence, Sediment. Geol., 155, 301–316, 2003.

Gralepois, M., Larrue, C., Wiering, M., Crabbé, A., Tapsell, S., Mees, H., Ek, K., Szwed, M.: Is flood defense changing in nature? Shifts in the flood defence strategy in six European countries, Ecology and Society, 21, 37, https://doi.org/10.5751/ES-08907-210437, 2016.

Grames, J., Prskawetz, A., Grass, D., Viglione, A., and Blöschl, G.: Modeling the interaction between flooding events and economic growth, Ecol. Econ., 129, 193–209, 2016.

Gray, C. L. and Mueller, V.: Natural disasters and population mobility in Bangladesh, PNAS, 109, 6000-6005, 2012.

Haque, C. E.: Human adjustments to river bank erosion hazard in the Jamuna floodplain, Bangladesh, Human Ecology, 16, 421-437, 1988.

Haque, C. E. and Zaman, M. Q.: Coping with riverbank erosion hazard and displacement in Bangladesh: Survival strategies and adjustments. Disasters, 13(4), 300–314, 1989.

Haque, C. E. and Zaman, M. Q.: Human responses to riverine hazards in Bangladesh: a proposal for sustainable floodplain development, World Development, 21, 93-107, 1993.

Hazarika, N., Das, A. K., and Borah, S. B.: Assessing land-use changes driven by river dynamics in chronically flood affected Upper Brahmaputra plains, India, using RS-GIS techniques, The Egyptian Journal of Remote Sensing and Space Science, 18, 107–118, https://doi.org/10.1016/j.ejrs.2015.02.001, 2015.

Hegger, D. L. T., Driessen, P. P. J., Wiering, M., Van Rijswick, H. F. M. W., Kundzewicz, Z. W, Matczak, P., Crabbé, A., Raadgever, G. T., Bakker, M. H. N., Priest, S. J., Larrue, C., Ek, K.: Toward more flood resilience: Is a diversification of flood risk management strategies the way forward?, Ecology and Society, 21, 52, https://doi.org/10.5751/ES-08854-210452, 2016.

Hino, M., Field, C. B., Mach, K. J.: Managed retreat as a response to natural hazard risk, Nature Climate Change, 7, 364, doi: 10.1038/NCLIMATE3252, 2017.

Hofer, T. and Messerli, B.: Floods in Bangladesh: History, dynamics and rethinking the role of the Himalayas, United Nations University Press, United Nations University, Tokyo, 1–468, 2006.

Hofer, T., & Messerli, B.: Floods in Bangladesh: history, dynamics and rethinking the role of the Himalayas. Ecology, 29, 254-283, 2006.

Hossain, M., Islam, A. T. M. A., Saha, S. K.: Floods in Bangladesh—An analysis of their nature and causes, In Floods in Bangladesh Recurrent Disaster and People's Survival, Universities Research Centre, Dhaka, Bangladesh, 1–21, 1987.

Hossain, M. Z.: Riverbank Erosion and Population Displacement: A Case of Kazipur in Pabna. Unpublished thesis, Jahangirnagar University, Savar, Dhaka, 1984.

Hornberger, G. M., Hess, D. J., and Gilligan, J.: Water conservation and hydrological transitions in cities in the United States, Water Resour. Res., 51, 4635–4649, https://doi.org/10.1002/2015WR016943, 2015.

Huq, H.: Flood action plan and NGO protests in Bangladesh: An assessment, in: Water governance and civil society responses in South Asia, edited by: Narayanan, N. C., Parasuraman, S., and Ariyabandu, R., New Delhi, India: Routlege, 2014.

Hutton, D. and Haque, E. E.: Human vulnerability, dislocation and resettlement: adaptation processes of river-bank erosion-induced displacees in Bangladesh. Disorder, 28, 41–62, 2004.

IECo: Master Plan. East Pakistan Water and Power Development Authority (EPWAPDA), 1964.

Indra, D.: Not just dis-placed and poor: How environmentally forced migrants in rural Bangladesh recreate space and place under trying conditions. Rethinking refuge and displacement, Selected papers of refugees and immigrants, Gozdziak, E.M.; Shandy, D., Eds.; American Anthropological Association, Washington, DC, USA, Volume VIII, 163-191, 2000.

IOM: Assessing the evidence: environment, climate change and migration in Bangladesh. International Organization for Migration, 2010.

Islam, M. A.: Human Adjustment to Cyclone Hazards: A Case Study of Char Jabbar, Natural Hazards Research, Working Paper No. 18, University of Toronto, Toronto, 1971.

Islam, N.: The Urban Poor in Bangladesh. Centre for Urban Studies, Dhaka, 1976.

Islam, M. S., Hasan, T., Chowdhury, M. S. I. R., Rahaman, M. H., Tusher, T. R.: Coping techniques of local people to flood and river erosion in char areas of Bangladesh, Journal of Environmental Science and Natural Resources, 5, 251-261, 2013.

Junk, W. J., Bayley, P. B., and Sparks, R. E.: The flood pulse concept in river–floodplain systems, in: Proceedings of the International Large River Symposium, 106, 110–127, 1989.

Joarder, M. A. M. and Miller, P. W.: Factors affecting whether environmental migration is temporary or permanent: Evidence from Bangladesh, Global Environmental Change, 23, 1511-1524, 2013.

Jongman, B., Winsemius, H. C., Aerts, J. C. J. H., de Perez, E. C., van Aalst, M. K., Kron, W., and Ward, P. J.: Declining vulnerability to river floods and the global benefits of adaptation, National Academy of Sciences, 1-10, doi: 10.1073/pnas.1414439112, 2015.

Kandasamy, J., Sounthararajah, D., Sivabalan, P., Chanan, A., Vigneswaran, S., and Sivapalan, M.: Socio-hydrologic drivers of the pendulum swing between agricultural development and environmental health: a case study from Murrumbidgee River basin, Australia, Hydrol. Earth Syst. Sci., 18, 1027–1041, https://doi.org/10.5194/hess-18-1027-2014, 2014.

Kates, R. W., Colten, C. E., Laska, S., and Leatherman, S. P.: Reconstruction of New Orleans after Hurricane Katrina?: A research perspective, P. Natl. Acad. Sci., 103, 14653–14660, https://doi.org/10.1073/pnas.0605726103, 2006.

Khan, A.A.M.: Rural-Urban Migration and Urbanization in Bangladesh. The Geographical Review 72(4): 379–94, 1982.

Khandker, S. R.: Coping with flood: role of institutions in Bangladesh, Agricultural Economics, 36, 169-180, 2007.

Kreibich, H., Di Baldassarre, G., Vorogushyn, S., Aerts, J. C. J. H., Apel, H., Aronica, G. T., Arnbjerg-Nielsen, K., Bouwer, L. M., Bubeck, P., Caloiero, T., Chinh, D. T., Cortès, M., Gain, A. K., Giampá, V., Kuhlicke, C., Kundzewicz, Z. W., and Llasat, M. C. B.: Adaptation to flood risk – results of international paired flood event studies, Earth's Future, 5, 953–965, https://doi.org/10.1002/2017EF000606, 2017.

Liao, K. H.: From flood control to flood adaptation: a case study on the Lower Green River Valley and the City of Kent in King County, Washington, Natural hazards, 71, 723-750, 2014.

Liu, Y., Tian, F., Hu, H., and Sivapalan, M.: Socio-hydrologic perspectives of the co-evolution of humans and water in the Tarim River basin, Western China: the Taiji-Tire model, Hydrol. Earth Syst. Sci., 18, 1289–1303, https://doi.org/10.5194/hess-18-1289-2014, 2014.

Logan, T. M., Guikema, S. D, Bricker, J. D.: Hard-adaptive measures can increase vulnerability to storm surge and tsunami hazards over time, Nature Sustainability, 1, 526–530, 2018.

Ludy, J., and Kondolf, G. M.: Flood risk perception in lands "protected" by 100-year levees, Natural hazards, 61, 829-842, 2012.

Lutz, W., Sanderson, W., and Scherbov, S.: Doubling of world population unlikely. Nature, 387, 803-805, 1997.

Macdonald, N., Chester, D., Sangster, H., Todd, B., and Hooke, J.: The significance of Gilbert F. White's 1945 paper 'Human adjustment to floods' in the development of risk and hazard management, Progress in Physical Geography, 36, 125-133, 2011.

Magliocca, N. R., Ellis, E. C., Allington, G. R. H., de Bremond, A., Dell'Angelo, J., Mertz, O., Messerli, P., Meyfroidt, P., Seppelt, R., and Verburg, P. H.: Closing global knowledge gaps: Producing generalized knowledge from case studies of social-ecological systems, Glob. Environ. Change, 50, 1–14, https://doi.org/10.1016/j.gloenvcha.2018.03.003, 2018.

Mamun, M. Z.: Awareness, Preparedness and Adjustment Measures of River-bank Erosion-prone People: A Case Study, Disasters, 20, 68-74, 2016.

Mård, J., Di Baldassarre, G. and Mazzoleni, M.: Nighttime light data reveal how flood protection shapes human proximity to rivers, Science Advances, 4, 1-7, doi: 10.1126/sciadv.aar5779, 2018.

Masozera, M., Bailey, M., and Kerchner, C.: Distribution of impacts of natural disasters across income groups: A case study of New Orleans, Ecol. Econ., 63, 299–306, 2007.

McCrum-Gardner, E.: Sample size and power calculations made simple, International Journal of Therapy and Rehabilitation, 17, 10-14, 2010.

McKinney, D. C.: Modeling water resources management at the basin level: Review and future directions, 6, 1999.

Mechler, R., Bouwer, L. M.: Understanding trends and projections of disaster losses and climate change: is vulnerability the missing link?, Climatic Change, 133, 23–35, 2015.

Mehta, V. K., Goswami, R., Kemp-Benedict, E., Muddu, S., and Malghan, D.: Metabolic urbanism and environmental justice: the water conundrum in Bangalore, India, Environmental Justice, 7, 130–137, https://doi.org/10.1089/env.2014.0021, 2014.

Merz, B., Elmer, F., and Thieken, A. H.: Significance of" high probability/low damage" versus" low probability/high damage" flood events, Natural Hazards and Earth System Sciences, 9, 1033-1046, 2009.

Milly, P. C. D., Wetherald, R. T., Dunne, K. A., and Delworth, T. L.: Increasing risk of great floods in a changing climate, Nature, 415, 514–517, 2002.

Milly, P. C., Betancourt, J., Falkenmark, M., Hirsch, R. M., Kundzewicz, Z. W., Lettenmaier, D. P., and Stouffer, R. J.: Stationarity is dead: Whither water management?, Science, 319, 573–574, https://doi.org/10.1126/science.1151915, 2008.

Mirza, M. M. Q.: Global warming and changes in the probability of occurrence of floods in Bangladesh and implications. Global environmental change, 12, 127-138, 2002.

Mirza, M. M. Q, Warrick, R. A., and Ericksen, N. J.: The implications of climate change on floods of the Ganges, Brahmaputra and Meghna rivers in Bangladesh, Climatic Change, 57, 287– 318, 2003.

Mondal, M. S., Jalal, M. R., Khan, M. S. A., Kumar, U., Rahman, R., and Huq, H.: Hydro-Meteorological Trends in Southwest Coastal Bangladesh: Perspectives of Climate Change and Human Interventions. American Journal of Climate Change, 2, 62-70, 2013.

Montanari, A., Young, G., Savenije, H. H. G., Hughes, D., Wagener, T., Ren, L. L., Koutsoyiannis, D., Cudennec, C., Toth, E., Grimaldi, S., Blöschl, G., Sivapalan, M., Beven, K., Gupta, H., Hipsey, M., Schaefli, B., Arheimer, B., Boegh, E., Schymanski, S. J., Di Baldassarre, G., Yu, B., Hubert, P., Huang, Y., Schumann, A., Post, D. A., Srinivasan, V., Harman, C., Thompson, S., Rogger, M., Viglione, A., McMillan, H., Characklis, G., Pang, Z., and Belyaev, V.: Panta Rhei – Everything Flows: Change in hydrology and society – The IAHS Scientific Decade 2013–2022, Hydrol. Sci. J., 58, 1256–1275, https://doi.org/10.1080/02626667.2013.809088, 2013.

Montz, B. E., and Tobin, G. A.: Livin' large with levees: Lessons learned and lost, Natural Hazards Review, 9, 150-157, 2008.

Moser, C. A., and Stuart, A.: An experimental study of quota sampling. Journal of the Royal Statistical Society, Series A (General), 116, 349-405, 1953.

Mostert, E.: An alternative approach for socio-hydrology: case study research, Hydrol. Earth Syst. Sci., 22, 317–329, https://doi.org/10.5194/hess-22-317-2018, 2018.

Nardi, F., Annis, A., Di Baldassarre, G., Vivoni, E. R., and Grimaldi, S.: GFPLAIN250m, a global high-resolution dataset of Earth's floodplains. Scientific data, 6, 180309, doi:10.1038/sdata.2018.309, 2019.

NDRCC: Daily Disaster Situation Report 18.9.2017: National disaster response co-ordination centre (NDRCC), Government of Bangladesh, 2017.

O'Connell, P. E. and O'Donnell, G.: Towards modelling flood protection investment as a coupled human and natural system, Hydrol. Earth Syst. Sci., 18, 155–171, https://doi.org/10.5194/hess-18-155-2014, 2014.

Ohl, C. and Tapsell, S.: Flooding and human health: the dangers posed are not always obvious, Brit. Med. J., 321, 1167–1168, 2000.

Opperman, J. J., Galloway G. E., Fargione J., Mount J. F., Richter B. D., and Secchi S.: Sustainable floodplains through large-scale reconnection to rivers, Science, 326, 1487–1488, 2009.

Pande, S. and Sivapalan, M.: Progress in socio-hydrology: a metaanalysis of challenges and opportunities, Wires Water, 4, 1–18, https://doi.org/10.1002/wat2.1193, 2017.

Paul, B. K.: Perception of and agricultural adjustment to floods in Jamuna floodplain, Bangladesh, Human Ecology, 12, 3-19, 1984.

Paul, B. K.: Flood research in Bangladesh in retrospect and prospect: a review, Geoforum, 28, 121-131, 1997.

Paul, S. K.; Routray, J. K.: Flood proneness and coping strategies: the experiences of two villages in Bangladesh, Disasters, 34, 489-508, 2010.

Peel, M. C. and Blöschl, G.: Hydrological modelling in a changing world, Prog. Phys. Geogr., 35, 249–261, https://doi.org/10.1177/0309133311402550, 2011.

Pelling, M.: The vulnerability of cities: Natural disasters and social resilience, Earthscan, London, 2003.

Penning-Rowsell, E. C., Sultana, P. and Thompson, P. M.: The last resort? Population movement in response to climate-related hazards in Bangladesh. Environmental Science and Policy, 27 (Supl 1), S44-S59. ISSN 1462-9011, 2012.

Rahman, A.: Human responses to natural hazards: The hope lies in social networking. Paper presented in the 23rd Bengal Studies Conference, University of Manitoba, Manitoba, Canada, June 9-11, 1989.

139

Rahman, L. M.: Present situation and future issues regarding river and hydrological data base in Bangladesh. In Proceedings of Second Experts Conference on River Information System, Sapporo, Japan, 31– 38, 1996.

Rahman, M. M.: Perceptions of flood risk and investment decisions in Bangladesh (Master's thesis). WSE-HERBD.17.10. UNESCO-IHE Institute for Water Education, Delft, the Netherlands, 2017.

Rahman, M. R.: Impact of riverbank erosion hazard in the Jamuna floodplain areas in Bangladesh, Journal of Science Foundation, 8, 55-65, 2013.

Rahman, T. M. A., Islam, S., Rahman, S. H.: Coping with flood and riverbank erosion caused by climate change using livelihood resources: a case study of Bangladesh, Climate and Development, 7, 185-191, 2015.

Rasid, H. and Mallik, A: Flood adaptations in Bangladesh: is the compartmentalization scheme compatible with indigenous adjustments of rice cropping to flood regimes?, Applied Geography, 15, 3-17, 1995.

Rasid, H. and Paul, B. K.: Flood problem in Bangladesh: is there an indigenous solution, Environmental Management, 11, 155-173, 1987.

RBIP: River Bank Improvement Program: Annex A, Vol 1, Morphology, Feasibility Report and Detailed Design Priority Reach, Bangladesh Water Development Board, Dhaka, 2015.

Reilly, A.C., Guikema, S.D, Zhu, L., Igusa, T.: Evolution of vulnerability of communities facing repeated hazards, PLoS ONE 12(9), e0182719, doi.org/10.1371/journal.pone.0182719, 2017.

Reuber, J., Schielen, R., and Barneveld, H. J.: Preparing a river for the future-The River Meuse in the year 2050, Floods, from Defence to Management, edited by: Van Alphen, J., van Beek, E., and Taal, M., Taylor & Francis Group, London, 687–692, 2005.

RMMRU: Coping with river bank erosion induced displacement. Policy Brief. Dhaka: University of Dhaka, RMMRU. Retrieved from http://www.rmmru.org, 2007.

Roy, M., Hanlon, J., Hulme, D.: Bangladesh Confronts Climate Change: Keeping Our Heads Above Water, 1st ed.; Anthem Press: London, UK and New York, USA, 1-173, 2016.

Sarker, M. H., Huque, I., Alam, M., and Koudstaal, R.: Rivers, chars and char dwellers of Bangladesh, Int. J. River Basin Manage., 1, 61–80, https://doi.org/10.1080/15715124.2003.9635193, 2003.

Sarker, M. H., Thorne, C. R., Aktar, M. N., and Ferdous, M. R.: Morpho-dynamics of the Brahmaputra–Jamuna River, Bangladesh Geomorphology, 215, 45-59. doi:10.1016/j.geomorph.2013.07.025, 2014.

Sarker, M. H: River Bank Improvement Program: Annex A, Vol 1, Morphology, Feasibility Report and Detailed Design Priority Reach, Bangladesh Water Development Board, Dhaka, Bangladesh, 1-127, 2015.

Schmuck, H.: "An Act of Allah": Religious Explanations for Floods in Bangladesh as Survival Strategy, International journal of mass emergencies and disasters, 18, 85-95, 2000.

Scolobig, A. and De Marchi, B.: Dilemmas in land use planning in flood prone areas, In: Samuels P, Huntington S, Allsop W, Harrop J (ed) Flood Risk Management: Research and Practice, Taylor and Francis Group, London, ISBN 978-0-415-48507-4, pp 204, 2009.

Shaw, R.: Living with floods in Bangladesh, Anthropology Today, 5, 11-13, 1989.

Shin, H. S., Hong, I., Kim, J. S., and Kim, K. H.: A study on variation of land-use in river area caused by levee construction, Journal of the Korea Academia-Industrial Cooperation Society, 15, 2419-2427, 2014.

Sivapalan, M.: Debates – Perspectives on socio-hydrology: Changing water systems and the "tyranny of small problems" – Sociohydrology, Water Resour. Res., 51, 4795–4805, https://doi.org/10.1002/2015WR017080, 2015.

Sivapalan, M. and Blöschl, G.: Time scale interactions and the coevolution of humans and water, Water Resour. Res., 51, 6988–7022, https://doi.org/10.1002/2015WR017896, 2015.

Sivapalan, M., Savenije, H. H. G., and Blöschl, G.: Sociohydrology: A new science of people and water, Hydrol. Process., 26, 1270–1276, https://doi.org/10.1002/hyp.8426, 2012.

Şorcaru, I. A.: Demographic Vulnerabilities in Tecuci Plain. Analele Universităţii din Oradea/The Annals of Oradea University, 23, 2013.

Sovacool, B. K.: Hard and soft paths for climate change adaptation, Climate Policy, 11, 1177–1183, 2011.

Srinivasan, V.: Reimagining the past – use of counterfactual trajectories in socio-hydrological modelling: the case of Chennai, India, Hydrol. Earth Syst. Sci., 19, 785–801, https://doi.org/10.5194/hess-19-785-2015, 2015.

Sultana, N., Rayhan, M. I.: Coping strategies with floods in Bangladesh: an empirical study, Natural hazards, 64, 1209-1218, 2012.

Sultana, P., Johnson, C., and Thompson, P.: The impact of major floods on flood risk policy evolution: Insights from Bangladesh, Int. J. River Basin Manage., 6, 339–348, https://doi.org/10.1080/15715124.2008.9635361, 2008.

Tanoue, M., Hirabayashi, Y., Ikeuchi, H.: Global-scale river flood vulnerability in the last 50 years, Nature Scientific Reports 6, article number: 36021, 2016.

Thompson, P., Tod, I.: Mitigating flood losses in the active floodplains of Bangladesh. Disaster Prevention and Management, An International Journal, 7, 113-123, 1998.

Tingsanchali, T. and Karim, M. F.: Flood hazard and risk analysis in the southwest region of Bangladesh, Hydrol. Process., 19, 2055–2069, https://doi.org/10.1002/hyp.5666, 2005.

Tobin, G. A.: The Levee Love Affair: A Stormy Relationship?, Water Resources Bulletin, 31, 359–367, doi: 10.1111/j.1752-1688.1995.tb04025.x, 1995.

Tongco, M. D. C.: Purposive sampling as a tool for informant selection, Ethnobotany Research and applications, 5, 147-158, 2007.

Tonn, G. L., and Guikema, S. D.: An agent-based model of evolving community flood risk, Risk Analysis, 38, 1258-1278, 2018.

Treuer, G., Koebele, E., Deslatte, A., Ernst, K., Garcia, M., and Manago, K.: A narrative method for analyzing transitions in urban water management: The case of the Miami-Dade Water and Sewer Department, Water Resour. Res., 53, 891–908, https://doi.org/10.1002/2016WR019658, 2017.

Troy, T. J., Pavao-Zuckerman, M., and Evans, T. P.: Debates – Perspectives on socio-hydrology: Socio-hydrologic modeling: Tradeoffs, hypothesis testing, and validation, Water Resour. Res., 51, 4806–4814, https://doi.org/10.1002/2015WR017046, 2015.

van Manen, S. E., and Brinkhuis, M.: Quantitative flood risk assessment for Polders. Reliability engineering & system safety, 90, 229-237, 2005.

van Staveren, M. F. and van Tatenhove, J. P. M.: Hydraulic engineering in the social-ecological delta: understanding the interplay between social, ecological, and technological systems in the Dutch delta by means of "delta trajectories", Ecol. Soc., 21, 8, https://doi.org/10.5751/ES-08168-210108, 2016.

van Staveren, M. F., van Tatenhove, J. P. M., and Warner, J. F.: The tenth dragon?: controlled seasonal flooding in long-term policy plans for the Vietnamese Mekong delta, Journal of Environmental Policy & Planning, 20, 267–281, https://doi.org/10.1080/1523908X.2017.1348287, 2017a.

van Staveren, M. F., Warner, J. F., and Khan, M. S. A.: Bringing in the tides. From closing down to opening up delta polders via Tidal River Management in the southwest delta of Bangladesh, Water Policy, 19, 147–164, https://doi.org/10.2166/wp.2016.029, 2017b.

Viglione, A., Di Baldassarre, G., Brandimarte, L., Kuil, L., Carr, G., Salinas, J. L., Scolobig, A., and Blöschl, G.: Insights from socio-hydrology modelling on dealing with flood risk – Roles of collective memory, risk-taking attitude and trust, J. Hydrol., 518, 71–82, https://doi.org/10.1016/j.jhydrol.2014.01.018, 2014.

Vis, M., Klijn, F., De Bruijn, K. M., and Van Buuren, M.: Resilience strategies for flood risk management in the Netherlands, Int. J. River Basin Manage., 1, 33–40, https://doi.org/10.1080/15715124.2003.9635190, 2003.

Walsham, M.: Assessing the evidence: environment, climate change and migration in Bangladesh. International Organization for Migration (IOM), Regional Office for South Asia, Dhaka, 2010.

Ward, P. J., Jongman, B., Aerts, J. C. J. H., Bates, P. D., Botzen, W. J. W., Loaiza, A. D., Hallegatte, S., Kind, J. M., Kwadijk, J., Scussolini, P. and Winsemius, H. C.: A global framework for future costs and benefits of river-flood protection in urban areas, Nature climate change, 7, 642–646, doi: 10.1038/NCLIMATE3350, 2017.

Wesselink, A.: Flood safety in the Netherlands: the Dutch political response to Hurricane Katrina, Technology in Society, 29, 239-247, 2007.

Wesselink, A., Warner, J., and Kok, M.: You gain some funding, you lose some freedom: the ironies of flood protection in Limburg (The Netherlands), Environmental Science and Policy, 30, 113–125, https://doi.org/10.1016/j.envsci.2012.10.018, 2013.

Wesselink, A., Warner, J., Abu Syed, S., Chan, F., Duc Tran, D., Huq, H., Huthoff, F., Le Thuy, N., Pinter, N., Van Staveren, M., Wester, P., and Zegwaard, A.: Trends in flood risk management in deltas around the world: are we going 'soft'?, International Journal of Water Governance, 3, 25-46, 2015.

Wesselink, A., Kooy, M., and Warner, J.: Socio-hydrology and hydrosocial analysis?: toward dialogues across disciplines, WIREs Water, 4, 1–14, https://doi.org/10.1002/wat2.1196, 2017.

Wesselink, A. J., Bijker, W. E., de Vriend, H. J., and Krol, M. S.: Dutch dealings with the Delta, Nature and Culture, 2, 188–209, https://doi.org/10.3167/nc2007.020203, 2007.

White, G. F.: Human Adjustment to Floods: Department of Geography Research, Paper No. 29, The University of Chicago, Chicago, 1945.

World Bank: The World Bank national accounts data, and OECD National Accounts data files. Available: https://data.worldbank.org/indicator/NY.GDP.PCAP.CD?locations=BD. (accessed on 16 January 2019), 2018a.

World Bank: The World Bank. Global Poverty Working Group. Available online: https://data.worldbank.org/country/bangladesh. (accessed on 16 January 2019), 2018b.

Yang, Y. C. E., Ray, P. A., Brown, C. M., Khalil, A. F., Yu, W. H.: Estimation of flood damage functions for river basin planning: a case study in Bangladesh, Natural Hazards, 75, 2773–2791, 2015.

Yasmin, T., Ahmed, K. M.: The comparative analysis of coping in two different vulnerable areas in Bangladesh, International Journal of Scientific & Technology Research, 2, 26-38, 2013.

Yu, D. J., Sangwan, N., Sung, K., Chen, X., and Merwade, V.: Incorporating institutions and collective action into a socio hydrological model of flood resilience, Water Resour. Res., 53, 1336–1353, https://doi.org/10.1002/2016WR019746, 2017.

Zaman, M. Q.: Rivers of life: living with floods in Bangladesh, Asian Survey, 33, 985-996, 1993.

APPENDICES

APPENDIX A: QUESTIONNAIRE FOR HOUSEHOLD SURVEY

Title: Socio-Hydrological dynamics in Bangladesh - Understanding the interaction between hydrological and social processes along the Jamuna floodplain

Questionnaire for household survey

Date: Sl. No:

GPS location/ mark in map:

Section A: General household information

1. Name of respondent:

2. Age:

3. Father's / husband's name:

4. Address:

5. Contact number (if any):

6. Distance of house from river (approximately):

7. Main occupation (household main income source):

   7a. Farmer: [] Large (land > 7.50 acres)   [] Medium (land 2.5-7.49 acres)   [] Small (land 0.5-2.49 acres)   [] Marginal (land 0.05-0.5 acres)   [] Share cropper

   7b. Fisher: [] Fishing in the rivers [] Fish culture in ponds [] Both

   7c. Businessman: (type)-

   7d. Service holder: (Institute/organization)-

   7e. Daily wage labor: [] Agricultural-labor [] Other (specify)

   7f. Housewife

   7g. Other (specify):

8. Distance of agricultural land from river (approximately):

9. Basic information about household

9a. Household Size:

9b. No. of earning member:

9c. Number of household members absent much of time here:

9d. Assets of the household: Land ownership

Type of Land	Area (dec)	Value (BDT/dec)
Agriculture		
Ponds		
Homestead		
Others		

9e. Other assets:

Assets	Number	Approximate value (BDT)
Housing materials/ construction type/size		
Livestock		
Others		

9f. Average annual income of the household (BDT):

Agriculture: Fisheries: Other sources list:

9g. Average monthly expenditure of the household (BDT):

9h. Housing Condition (building material):

10. How long are you living here? .... years.

11. Information about origin? [] Born here   [] Immigrated from another place

12. Reason of immigration (Only applicable for immigrated people from another place):

12a. From where did you come here?

12 b. What is the reason/s of your immigration?

12c. Before immigration, did you know if this location was exposed to flooding or not?
[] Yes [] No

12d. Before immigration, did you know if this location was exposed to riverbank erosion or not? [] Yes [] No

Section B: Flood experiences of the household

13. Information about flood:

13a. Have you seen floods in your area before? [] Yes [] No

13b. If yes, please specify:

Years	Months	Flood depth (feet) in your homestead	Flood depth (feet) in your agriculture land	Flood duration (days)

13c. From where does the flood come to your home or agricultural land?

[] Heavy rainfall [] From Jamuna river [] Other rivers/channels (please specify):

13d. Agriculture cropping pattern vulnerable to flood:

Name of crop/s: Stage: Seedling / Vegetative/ Flowering/ Ripening

13e. Did you change your agricultural copping pattern due to flooding? [] Yes [] No

If yes, then how many times and how?

13f. Did you change your agriculture land use pattern due to flooding? [] Yes [] No

Years	Permanently/ Temporarily	Land from- to	Changed area (dec)	Reason for changing

13g. List of household's assets damaged in these flood events (same year as Q13b) and their value:

Household assets	Value (BDT)					
Years						
Agriculture crops						
Farming tools (e.g. pump)						
Fisheries						
Livestock						
House and HH materials						
Others:						

14. How did you recover from these damages? What were the strategies?

15. In your opinion, what is main cause of these floods?

      [] Excessive monsoon rain for several days

      [] Increased discharge of the nearby rivers and channels

      [] Rapid sedimentation of riverbeds

      [] Sea level rise

      [] Others (please specify; if, necessary explain)

16. What are the effects of these floods you have experienced?

      [a] Totally displaced from home,

      [b] Partially (temporarily) displaced from home,

      [c] Not displaced but the home was submerged under flood water,

      [d] Lost all field Crops,

[e] Lost all fisheries,

[f] Loss of daily income,

[g] Suffered from diarrhea/stomach flu/pneumonia/cold/fever/skin disease/snake bite/death,

[h] Others (please specify; if, necessary explain)

Years	Codes	Others (please specify)

17. Coping strategies of flooding: For homestead:

[a] Raising homestead   [b] Raised platform for human

[c] Raised platform for food, fuel, water   [d] Moving to flood shelters/ safer places

[e] Raised platform for poultry and livestock [f] Changing eating behavior

[g] Use purifying tablets for drinking water   [h] Portable stove for cooking

[i] Transferable construction materials of house

[j] Plantation/ iron sheet/ bamboo around house to prevent erosion

[k] Others (please specify):

Years	Codes	Others (please specify)

For agriculture, aquaculture and other economic activities and income generation:

[a] Flood resistant crops (crops:)

[b] Planting a short duration crop after floods

[c] Keep fallow during flood season

[d] Grow vegetable in raised land/ platform

[e] Protect fish pond with net

[f] Alternative source of income during flood

[g] Selling women jewelry

[h] Borrowing from others

[i] Mortgaging and selling land and other productive assets.

[j] Others (please specify):

Years	Codes	Others (please specify)

18. Did you change your occupation due to flood? [] Yes   [] No

Years	Permanently/ Temporarily	Previous occupation	Changed occupation

19. Has your income level decreased due to flood? [] Yes   [] No

   If yes then, [] Permanently   [] Temporarily

   Mention monthly amount and year?

20. Has your cost of living increased due to flood? [] Yes   [] No

   If yes then, [] Permanently   [] Temporarily

   Mention monthly amount and year?

21. Is there scarcity of job opportunity due to flood? [] Yes   [] No

   If yes, mention how does this arise?

153

22. Is there any new job opportunity due to flood? [] Yes [] No

If yes, mention how does this arise?

23. What is your perception about controlling flood with structure like embankment or others?

[] Good for society, it saves lives and damages

[] Not good for society. (Why? Explain:)

[] Others (please specify)

24. The presence of an irrigation project (with its bunds etc) would affect the pattern of flooding?

[] Yes [] No

Please clarify:

25. Are you thinking of migration to another place for better opportunity? [] Yes [] No

If yes, mention why and where?

[] Temporary migration for few months
(explain:)

[] Permanent migration with family
(explain:)

26. Do you know anyone (neighbor, family friends) who had migrated from this area?
[] Yes [] No

If yes, mention why, which year and where?

[] Temporary migration for few months
(explain:)

[] Permanent migration with family
(explain:)

Please specify if there are multiple possibilities:

27. Did you receive any early warning message about the flooding? [] Yes [] No

Years	how many days before of the event	from whom	how/medium

28. What measures were taken by the government against flooding?

29. Did you do any special flood response activity for the society? If yes, list them below.

Section C: River bank erosion experiences of the household

30. Information about river bank erosion: *(if char erosion then mention)*

    30a. Have you ever experienced with river bank erosion? [] Yes [] No

    30b. Did you migrate due to river bank erosion? If yes, when and how many times in your life?

Years	Old place	New place	Distance (km)	Reasons

    30c. Have you lost your lands due to river bank erosion? [] Yes [] No

Years (with occurring month)	Type of lands	Area (ha)	Value (BDT)

    If no, then do you think that you will lose your lands in future? [] Yes [] No

    30d. Have you lost your homestead and move home due to river bank erosion?

[] Yes [] No

Years (with occurring month)	Area (ha)	Value (BDT)	Homestead assets lost	Value (BDT)

If no, then do you think that you will be forced to move home in future? [] Yes [] No

30e. Did you change your agricultural cropping pattern due to river bank erosion? [] Yes [] No

If yes, then how?

31. How did you recover from these damages listed in Q309c, Q30d?

32. In your opinion, what is main cause of river bank erosion?

33. Did you receive any assistance during riverbank erosion? [] Yes [] No

If yes, from whom and how?

34. In your opinion and experience, has the Jamuna embankment helped to protect people from river bank erosion?

35. Coping with riverbank erosion:

For homestead:

For agriculture, fishing and other economic activities and income generation:

36. Did you change your occupation due to riverbank erosion? [] Yes [] No

If yes then, [] Permanently [] Temporarily

Mention the previous occupation(s) and year of change(s)?

37. Has your income level changes due to riverbank erosion? [] Yes [] No

If yes then, [] Fell [] Same [] Rose and [] Permanently [] Temporarily

Mention monthly amount and year?

38. Has your cost of living changed due to riverbank erosion? [] Yes   [] No

   If yes then, [] Fell [] Same [] Rose and [] Permanently   [] Temporarily

   Mention monthly amount and year?

39. Did you receive any early warning message about the riverbank erosion?

   [] Yes   [] No

Years	how many days before of the event	from whom	how/medium

40. What measures taken by the government against riverbank erosion?

41. Do you have any additional comments?

APPENDIX B: AGENDA FOR FOCUS GROUP DISCUSSION WITH LIVELIHOOD GROUPS

Title: Socio-Hydrological dynamics in Bangladesh - Understanding the interaction between hydrological and social processes along the Jamuna floodplain

Agenda for Focus Group Discussion with livelihood groups

- Introduction: discus about the research

Flooding

- How flood is affecting the livelihood group and what are the copping strategies?
- What are the main reasons of flooding in your area?
- Is the livelihood group has the tendency or bound to change their occupation due to flooding?
- Is the livelihood group has the tendency or bound to migrate due to flooding?
- Did your neighbors migrate due to flooding?
- Do you think any human activity has influence on flooding or increase of flooding?
- Do you receive any flood forecast form any where?
- What strategies government taken against flooding?

River bank erosion

- How riverbank erosion is affecting the livelihood group and what are the copping strategies?
- Is the livelihood group has the tendency or bound to change their occupation due to riverbank erosion?
- Is the livelihood group has any tendency to migrate to another places due to riverbank erosion?
- Do you receive any forecast regarding river bank erosion?

- Do you think any human activity has influence on riverbank erosion?
- What are the initiatives from the government or local administration to mitigate the impacts?

APPENDIX C: STATISTICAL ANALYSIS TO TEST DIFFERENCES BETWEEN THE SHSs

C1: Source of flooding

H_0: Is there a significant difference in sources of flooding by the socio-hydrological spaces?

To test the above hypothesis, we performed Chi-square tests. We find that there is a statistically significant difference in sources of flooding between the different socio-hydrological spaces. SHS1 is significantly more often flooded from other rivers and excessive rainfall than SHS2, while SHS2 is significantly more often flooded from the Jamuna River. SHS3 is significantly more often flooded from other rivers than SHS2, while SHS2 is significantly more often flooded from the Jamuna River. SHS1 is significantly more often flooded from excessive rainfall than SHS3, while SHS3 is significantly more often flooded from the Jamuna River. Detail analysis are shown below.

Observed	Socio-hydrological spaces			Observed	Socio-hydrological spaces		
Sources	SHS1	SHS2	Total	Sources	SHS1	SHS3	Total
Jamuna	233	298	531	Jamuna	233	267	500
Non-Jamuna	53	0	53	Non-Jamuna	53	12	65
Total	286	298	584	Total	286	279	565
Expected	Socio-hydrological spaces			Expected	Socio-hydrological spaces		
Sources	SHS1	SHS2	Total	Sources	SHS1	SHS3	Total
Jamuna	260.04	270.96	531	Jamuna	253.10	246.90	500
Non-Jamuna	25.96	27.04	53	Non-Jamuna	32.90	32.10	65
Total	286	298	584	Total	286	279	565
p =	6.53E-15			p =	1.16E-07		

Observed	Socio-hydrological spaces		
Sources	SHS2	SHS3	Total
Jamuna	298	267	565
Non-Jamuna	0	12	12
Total	298	279	577

Expected	Socio-hydrological spaces		
Sources	SHS2	SHS3	Total
Jamuna	291.80	273.20	565
Non-Jamuna	6.20	5.80	12
Total	298	279	577

p =	2.97E-04		

Observed	Socio-hydrological spaces		
Sources	SHS1	SHS2	Total
Excessive rainfall	28	0	28
Non-Excessive rainfall	258	298	556
Total	286	298	584

Expected	Socio-hydrological spaces		
Sources	SHS1	SHS2	Total
Excessive rainfall	13.71	14.29	28
Non-Excessive rainfall	272.29	283.71	556
Total	286	298	584

p =	3.10E-08		

Observed	Socio-hydrological spaces		
Sources	SHS1	SHS3	Total
Excessive rainfall	28	2	30
Non-Excessive rainfall	258	277	535
Total	286	279	565

Expected	Socio-hydrological spaces		
Sources	SHS1	SHS3	Total
Excessive rainfall	15.19	14.81	30
Non-Excessive rainfall	270.81	264.19	535
Total	286	279	565

p =	1.52E-06		

Observed	Socio-hydrological spaces		
Sources	SHS2	SHS3	Total
Excessive rainfall	0	2	2
Non-Excessive rainfall	298	277	575
Total	298	279	577

Expected	Socio-hydrological spaces		
Sources	SHS2	SHS3	Total
Excessive rainfall	1.03	0.97	2
Non-Excessive rainfall	296.97	278.03	575
Total	298	279	577

p =	0.14316		

161

Observed	Socio-hydrological spaces		
Sources	SHS1	SHS2	Total
Other rivers	120	0	120
Non-other rivers	166	298	464
Total	286	298	584

Expected	Socio-hydrological spaces		
Sources	SHS1	SHS2	Total
Other rivers	58.77	61.23	120
Non-other rivers	227.23	236.77	464
Total	286	298	584

p =	4.25E-36		

Observed	Socio-hydrological spaces		
Sources	SHS1	SHS3	Total
Other rivers	120	106	226
Non-other rivers	166	173	339
Total	286	279	565

Expected	Socio-hydrological spaces		
Sources	SHS1	SHS3	Total
Other rivers	114.40	111.60	226
Non-other rivers	171.60	167.40	339
Total	286	279	565

p =	0.33611		

Observed	Socio-hydrological spaces		
Sources	SHS2	SHS3	Total
Other rivers	0	106	106
Non-other rivers	298	173	471
Total	298	279	577

Expected	Socio-hydrological spaces		
Sources	SHS2	SHS3	Total
Other rivers	54.75	51.25	106
Non-other rivers	243.25	227.75	471
Total	298	279	577

p =	5.13E-32		

C2: Flood occurrence

H₀: Is there a significant difference in flood occurrences between the socio-hydrological spaces?

H_0: *Is there a significant difference in flood occurrences between the socio-hydrological spaces?*

To test the above hypothesis, we performed Chi-square tests. We find that there is a statistically significant difference in number of floods between the socio-hydrological spaces. There is no significant difference in number of floods between SHS1 and SHS3. Detail analysis are shown below.

Observed Occurances	Socio-hydrological spaces			Observed Occurances	Socio-hydrological spaces		
	SHS1	SHS2	Total		SHS1	SHS3	Total
Flooded	33	54	87	Flooded	33	30	63
Non-flooded	21	0	21	Non-flooded	21	24	45
Total	54	54	108	Total	54	54	108

Expected Sources	Socio-hydrological spaces			Expected Sources	Socio-hydrological spaces		
	SHS1	SHS2	Total		SHS1	SHS3	Total
Flooded	43.50	43.50	87	Flooded	31.50	31.50	63
Non-flooded	10.50	10.50	21	Non-flooded	22.50	22.50	45
Total	54	54	108	Total	54	54	108

p =	3.29E-07			p =	0.558185		

Observed Occurances	Socio-hydrological spaces		
	SHS2	SHS3	Total
Flooded	54	30	84
Non-flooded	0	24	24
Total	54	54	108

Expected Sources	Socio-hydrological spaces		
	SHS2	SHS3	Total
Flooded	42.00	42.00	84
Non-flooded	12.00	12.00	24
Total	54	54	108

p =	2.78E-08		

C3: Average flood damage

H₀: Is there a significant difference in average flood damages between the hydrological spaces?

To test the above hypothesis, we use a single-factor analysis of variance (ANOVA). We find that there is no statistically significant difference in average flood damages between the socio-hydrological spaces (with α=0.05). Detail analysis is shown below.

Anova: Single Factor

SUMMARY

Groups	Count	Sum	Average	Variance
SHS1	286	165244.7	577.7786	432246.2
SHS2	298	172707.1	579.554	432091
SHS3	279	140711	504.3406	475834.1

ANOVA

Source of Variation	SS	df	MS	F	P-value	F crit
Between Groups	1043972	2	521986	1.169631	0.310975	3.006192
Within Groups	3.84E+08	860	446282.6			
Total	3.85E+08	862				

C4: Experienced with river bank erosion

H_0: Is there a significant difference in experience with river bank erosion between the Socio-hydrological spaces?

To test the above hypothesis, we performed Chi-square tests. We find that there is a statistically significant difference in experience with river bank erosion between the socio-hydrological spaces. Detail analysis are shown below.

Observed Erosion	Socio-hydrological spaces SHS1	SHS2	Total	Observed Erosion	Socio-hydrological spaces SHS1	SHS3	Total
Eroded	44	243	287	Eroded	44	142	186
Non-eroded	242	55	297	Non-eroded	242	137	379
Total	286	298	584	Total	286	279	565

Expected Erosion	Socio-hydrological spaces SHS1	SHS2	Total	Expected Erosion	Socio-hydrological spaces SHS1	SHS3	Total
Eroded	140.55	146.45	287	Eroded	94.15	91.85	186
Non-eroded	145.45	151.55	297	Non-eroded	191.85	187.15	379
Total	286	298	584	Total	286	279	565

p =	1.57E-57			p =	2.69E-19		

Observed Erosion	Socio-hydrological spaces SHS2	SHS3	Total
Eroded	243	142	385
Non-eroded	55	137	192
Total	298	279	577

Expected Erosion	Socio-hydrological spaces SHS2	SHS3	Total
Eroded	198.84	186.16	385
Non-eroded	99.16	92.84	192
Total	298	279	577

p =	5.83E-15		

C5: Average total wealth and income

H$_0$: Is there a significant difference in average total wealth between the socio-hydrological spaces?

To test the above hypothesis, we use a single-factor analysis of variance (ANOVA). We find that there exists a statistically significant difference in average total wealth between the socio-hydrological spaces (with α=0.05). Detail analysis is shown below.

Anova: Single Factor

SUMMARY

Groups	Count	Sum	Average	Variance
SHS1	286	9863089	34486.32	2.05E+09
SHS2	298	1782068	5980.095	98264341
SHS3	279	4923220	17645.95	6.4E+08

ANOVA

Source of Variation	SS	df	MS	F	P-value	F crit
Between Groups	1.19584E+11	2	5.98E+10	64.99275	5.18E-27	3.006192
Within Groups	7.91183E+11	860	9.2E+08			
Total	9.10767E+11	862				

H_0: Is there a significant difference in average annual income in between the socio-hydrological spaces?

To test the above hypothesis, we use a single-factor analysis of variance (ANOVA). We find that there exists a statistically significant difference in average annual income between the socio-hydrological spaces (with $\alpha=0.05$). Detail analysis is shown below.

Anova: Single Factor

SUMMARY

Groups	Count	Sum	Average	Variance
SHS1	286	361806.3	1265.057	539348.3
SHS2	298	289100	970.1342	345594.9
SHS3	279	345125	1237.007	3666301

ANOVA

Source of Variation	SS	df	MS	F	P-value	F crit
Between Groups	15524131	2	7762066	5.233178	0.005508	3.006192
Within Groups	1.28E+09	860	1483241			
Total	1.29E+09	862				

C6: Migration

H₀: Is there a significant difference in migration/relocation between the socio-hydrological spaces?

To test the above hypothesis, we performed Chi-square tests. We find that there is a statistically significant difference in migration/relocation between the socio-hydrological spaces. Detail analysis are shown below.

Observed	Socio-hydrological spaces			Observed	Socio-hydrological spaces		
Migration	SHS1	SHS2	Total	Migration	SHS1	SHS3	Total
Migrated	71	238	309	Migrated	71	116	187
Non-migrated	215	60	275	Non-migrated	215	163	378
Total	286	298	584	Total	286	279	565
Expected	Socio-hydrological spaces			Expected	Socio-hydrological spaces		
Erosion	SHS1	SHS2	Total	Erosion	SHS1	SHS3	Total
Eroded	151.33	157.67	309	Eroded	94.66	92.34	187
Non-eroded	134.67	140.33	275	Non-eroded	191.34	186.66	378
Total	286	298	584	Total	286	279	565
p =	1.75E-40			p =	2.33E-05		

Observed	Socio-hydrological spaces		
Migration	SHS2	SHS3	Total
Migrated	238	116	354
Non-migrated	60	163	223
Total	298	279	577
Expected	Socio-hydrological spaces		
Erosion	SHS2	SHS3	Total
Eroded	182.83	171.17	354
Non-eroded	115.17	107.83	223
Total	298	279	577
p =	3.77E-21		

C7: Homestead types

H₀: Is there a significant difference between the homestead types?

To test the above hypothesis, we performed Chi-square tests. We find that there are significantly more houses made of brick and tin and well-constructed buildings using modern masonry materials in SHS1, than the two other spaces. SHS2 has significantly more houses made of straw, while SHS3 has significantly less houses made of earthen floor, wood, paddy straw and bamboo mats. Detail analysis are shown below.

Observed	Socio-hydrological spaces			Observed	Socio-hydrological spaces		
Homestead types	SHS1	SHS2	Total	Homestead types	SHS1	SHS3	Total
Earthern	219	226	445	Earthern	219	247	466
Non-earthern	67	72	139	Non-earthern	67	32	99
Total	286	298	584	Total	286	279	565

Expected	Socio-hydrological spaces			Expected	Socio-hydrological spaces		
Homestead types	SHS1	SHS2	Total	Homestead types	SHS1	SHS3	Total
Earthern	217.93	227.07	445	Earthern	235.89	230.11	466
Non-earthern	68.07	70.93	139	Non-earthern	50.11	48.89	99
Total	286	298	584	Total	286	279	565

p =	0.834952			p =	0.000186		

Observed	Socio-hydrological spaces		
Homestead types	SHS2	SHS3	Total
Earthern	226	247	473
Non-earthern	72	32	104
Total	298	279	577

Expected	Socio-hydrological spaces		
Homestead types	SHS2	SHS3	Total
Earthern	244.29	228.71	473
Non-earthern	53.71	50.29	104
Total	298	279	577

p =	7.39E-05		

Observed	Socio-hydrological spaces		
Homestead types	SHS1	SHS2	Total
Well-constructed	7	0	7
Non-well-constructed	279	298	577
Total	286	298	584

Expected	Socio-hydrological spaces		
Homestead types	SHS1	SHS2	Total
Well-constructed	3.43	3.57	7
Non-well-constructed	282.57	294.43	577
Total	286	298	584

p =	0.006587		

Observed	Socio-hydrological spaces		
Homestead types	SHS1	SHS3	Total
Well-constructed	7	1	8
Non-well-constructed	279	278	557
Total	286	279	565

Expected	Socio-hydrological spaces		
Homestead types	SHS1	SHS3	Total
Well-constructed	4.05	3.95	8
Non-well-constructed	281.95	275.05	557
Total	286	279	565

p =	0.035609		

Observed	Socio-hydrological spaces		
Homestead types	SHS2	SHS3	Total
Well-constructed	0	1	1
Non-well-constructed	298	278	576
Total	298	279	577

Expected	Socio-hydrological spaces		
Homestead types	SHS2	SHS3	Total
Well-constructed	0.52	0.48	1
Non-well-constructed	297.48	278.52	576
Total	298	279	577

p =	0.300956		

Observed	Socio-hydrological spaces		
Homestead types	SHS1	SHS2	Total
Straw	25	64	89
Non-straw	261	234	495
Total	286	298	584

Expected	Socio-hydrological spaces		
Homestead types	SHS1	SHS2	Total
Straw	43.59	45.41	89
Non-straw	242.41	252.59	495
Total	286	298	584

p =	1.86E-05		

Observed	Socio-hydrological spaces		
Homestead types	SHS1	SHS3	Total
Straw	25	27	52
Non-straw	261	252	513
Total	286	279	565

Expected	Socio-hydrological spaces		
Homestead types	SHS1	SHS3	Total
Straw	26.32	25.68	52
Non-straw	259.68	253.32	513
Total	286	279	565

p =	0.700343		

Observed	Socio-hydrological spaces		
Homestead types	SHS2	SHS3	Total
Straw	64	27	91
Non-straw	234	252	486
Total	298	279	577

Expected	Socio-hydrological spaces		
Homestead types	SHS2	SHS3	Total
Straw	47.00	44.00	91
Non-straw	251.00	235.00	486
Total	298	279	577

p =	0.000102

Observed	Socio-hydrological spaces		
Homestead types	SHS1	SHS2	Total
Brick and tin	35	8	43
Non-brick and tin	251	290	541
Total	286	298	584

Expected	Socio-hydrological spaces		
Homestead types	SHS1	SHS2	Total
Brick and tin	21.06	21.94	43
Non-brick and tin	264.94	276.06	541
Total	286	298	584

p =	9.92E-06

Observed	Socio-hydrological spaces		
Homestead types	SHS1	SHS3	Total
Brick and tin	35	4	39
Non-brick and tin	251	275	526
Total	286	279	565

Expected	Socio-hydrological spaces		
Homestead types	SHS1	SHS3	Total
Brick and tin	19.74	19.26	39
Non-brick and tin	266.26	259.74	526
Total	286	279	565

p =	4.09E-07

Observed	Socio-hydrological spaces		
Homestead types	SHS2	SHS3	Total
Brick and tin	8	4	12
Non-brick and tin	290	275	565
Total	298	279	577

Expected	Socio-hydrological spaces		
Homestead types	SHS2	SHS3	Total
Brick and tin	6.20	5.80	12
Non-brick and tin	291.80	273.20	565
Total	298	279	577

p =	0.29271

Appendix D: Detail data and method of the levee effect analysis

D1: Primary data for the two rural areas (area 1 and area 2)

A cross-sectional method was used to gather these primary data. Cross-sectional research (Barnett et al., 2012) involves using different groups of people, both male and female (farmer, fisherman, day-labour, service holder etc.) who differ in the variables of interest but share other characteristics, such as socio-economic status and ethnicity. Due to the rural character of the area, most respondents were farmers. An age bias was introduced to collect historical information on flooding, riverbank erosion, livelihood etc. The household surveys were implemented with a combination of purposive sampling and quota sampling. In purposive sampling (Tongco, 2007) individuals are selected because they meet specific criteria (e.g., farmer, fisherman, day labour etc.); the quota sampling method (Moser and Stuart, 1953) selects a specific number of respondents with particular qualities (like farmer's age should be 40 or above). The Raosoft sample size calculator (McCrum-Gardner, 2010) was used to determine the required sample size for the surveys by union (the lowest administrative unit of Bangladesh government). In this calculator the researcher enters values including acceptable margin of error, response distribution, confidence level and size of the population that is to be surveyed. We accepted a 5% margin of error with 95% confidence level to determine the sample size, which is 1% households (560 household surveys) of the two rural study areas in Gaibandha (area 1) and Jamalpur (area 2). We performed ANOVA test ($p<0.05$) on the results presented below, which showed that all results are significantly different except where this is explicitly mentioned.

D2: Primary data for urban areas (area 3 and area 4)

Each meeting with the investor count as one sample for data analysis. At Gaibandha town 31 sample of investment and 34 sample at Sirajganj town were collected from the field investigation. All the samples are private investment. The questions used to elicit flood risk perceptions were designed based on the different variables, such as reasons for investment decision, awareness, worry, probability, impact etc.

A question was asked to the respondents about reasons for taking that specific investment. Risk awareness of flood is measured by two questions. The first relates to the current flood risk and ask to the respondents whether he/she knows that the area is in flood affected area or not. The second question relates to the financial analysis and ask to the respondent whether he/she perform any financial analysis or not and if they analysed then what type of risk or advantages they take into account related to flood risk. Another question was asked to the respondent about the worry about the flood risk that whether he/she wants to leave or quit the investment due to the effect of flooding or not. A question was asked to the respondent how often they are affected by flood. A table was added to the questionnaire about the impact of past flood on respondent's investment. The question format enables to locate the respondent's perception about flood risk and find out the relationship between flood risk perception and investment decision.

Collected data were classified on the basis of categories that were emerge from the data. The qualitative data were quantified through a coding mechanism. After completing the data cleaning process statistical analysis were conducted. Both univariate and bivariate analysis were conducted, for example to investigate the existence of correlation.

D3: Secondary data

We have collected the population data for Sirajganj and Gaibandha towns from 1901 to 1961 (CPP, 1964) and detailed population data for the whole study area from 1974 to 2011 (BBS, 1974; BBS, 1986; BBS, 1994; BBS, 2005, BBS, 2013 and BBS, 2014a, b). We have extracted road network for 1943 from aerial maps of 1943 and the road network for 2014 from the satellite image. Land use and Land cover classification was carried out using optical images for this study as these images were of high spectral resolution (7 bands for Landsat 4/5 and Landsat ETM, and 11 bands for Landsat 8) and suitable for land use and land cover mapping. Due to lack of availability of good spatial resolution remote sensing data prior to 1989, we could analyse the settlement areas only for the years 1989, 1997, 2003, 2010 and 2014, which shows the expansion during last three decades. All images were geo-rectified into "Bangladesh Transverse Mercator" (BTM) projection. The settlement data of 2003 was taken from the vector layer data and prepared using Cartosat Panchromatic images of 2.5m of spatial resolution. The settlement data of 2014 was taken from vector data which was digitized from multispectral RapidEye (5m of spatial resolution) images. These vector data were converted into raster format and used for finalizing the classification of the land use of 2014.

APPENDIX E: QUESTIONNAIRE FOR PRIMARY DATA COLLECTION FOR URBAN AREAS

Questionaries' for Primary data collection for the two urban areas (area 3 and 4)

Date: Serial No.:

GPS Location/Mark in map:

A. General Information:

1. Name:

2. Age:

3. Company name:

4. Role in company

   [ ] Owner

   [ ] Manager

   [ ] other:

5. Address:

6. Mobile No.:

B. Investment Information:

7. Type of investment (industry):

   [ ] Production type – factories

   [ ] Investment type – Hotel, Apartment

   [ ] Others:

8. Amount of land where you have invested?

9. Property right on land?

175

[ ] Owned

[ ] Leased

[ ] Loan

[ ] Others:

10. Amount of investment (BDT):

11. When you have started this? (Year):

12. Why you have invested here?

   [ ] Inherited

   [ ] Encouraged by previous investment decision

   [ ] Perceived advantages [explain]

   [ ] Low price of land

   [ ] Availability of raw materials

   [ ] Others:

Details:

C. Flood Risk Management, Cost-Benefit analysis:

13. Did you know that this is a flood affected area before investment? [ ] Yes [ ] No

If yes then why you invest?

14. Do you want to invest, if you know that there is risk of flooding? [ ] Yes [ ] No

If yes, why?

15. Did you perform any financial analysis related to flood risk before investment?

[ ] Yes [ ] No

If yes then what type of financial cost did you consider related to flood risk?

Risk:

Advantages:

16. Did you take any other investment decision before this in this area / elsewhere?

[ ] Yes [ ] No

17. If yes then, what was the result of that investment?

[ ] Failed

[ ] Successful

Explain:

D. Flood Characteristics:

18. How much distance from the river (approximately):

19. What is the frequency of flooding in this area?

20. Did you see flood in this area: [ ] Yes   [ ] No

If yes then,

Year				
Depth of flood (feet)				
Duration of flood (days)				
Amount of damage (BDT)				
Amount of damage (equipment) (BDT)				
Amount of damage (Building & Furniture) (BDT)				
Other damage (BDT)				

21. In your opinion, what is the main cause of the flood?

22. Brahmaputra Right Embankment (BRE) was constructed to protect flooding, what do you think about this?

[ ] Feeling safe to invest, reduce damage due to flooding.

[ ] Flood damages are same as unprotected area

[ ] Never considered the effect of flood   [ ] Others:

23. This year BRE was breached once again (only Gaibandha), does it change your perception about flood risk?

[ ] Yes   [ ] No.

If yes, what type of change?

24. Does it change your perceptions to invest? [ ] Yes   [ ] No.

If yes, how?

25. Do you want to leave or quit this investment? [ ] Yes [ ] No

26. Do you know anyone who wanted to invest but decided not to because of flood risk?

[ ] Yes [ ] No

If yes, could we have contact details?

APPENDIX F: DETAIL DATA AND METHOD OF THE COSTS OF LIVING WITH FLOODS

F1: Method of collecting primary data and analyses

A cross-sectional method was used to gather these primary data. Cross-sectional research involves using different groups of people, both male and female (farmer, fisherman, day-labour, service holder etc.) who differ in the variables of interest but share other characteristics, such as socio-economic status and ethnicity. We aimed to collect approximately the same number of surveys in each of the three SHS. Due to the rural character of the area, most respondents were farmers. An age bias was introduced to collect historical information on flooding, riverbank erosion, livelihood etc. The household surveys were implemented with a combination of purposive sampling and quota sampling. In purposive sampling individuals are selected because they meet specific criteria (e.g., farmer, fisherman, day labour etc.); the quota sampling method selects a specific number of respondents with particular qualities (like farmer's age should be 40 or above). The Raosoft sample size calculator was used to determine the required sample size for the surveys by union (the lowest administrative unit of Bangladesh government). In this calculator the researcher enters values including acceptable margin of error, response distribution, confidence level and size of the population that is to be surveyed. We accepted a 5% margin of error with 95% confidence level to determine the sample size, which is 1% households (863 household surveys) of the study area. In addition, we performed 12 focus group discussions in the study area, four meetings in different unions in each SHS. About 20 participants were present in each of the meetings. Participants were selected based on occupation and location of the households, guaranteeing a uniform spread over the union area. The topics of the discussions were: how is flooding affecting livelihoods; what household coping strategies are used in relation to flooding, for example changing occupation or raising homesteads; migration patterns; community interventions against flooding; river bank erosion and household coping strategies; community interventions against riverbank erosion; governmental initiatives against flooding and riverbank erosion etc.

F2: Secondary data

Because of paucity of secondary socio-economic data, we have used only for the years 2003 and 2010 in district level. We have collected detailed demographic data (up to village/Mauza level) from 1961 to the latest published census of 2011. The river water level data (1960-2015) was collected to correlate with the primary flooding data collected during household surveys. We only could collect Radarsat images for the years 1998 – 2004 and they were used to compere the flooded area with flood information of the households. All images were geo-rectified into "Bangladesh Transverse Mercator" (BTM) projection before analysing.

APPENDIX G: STATISTICAL ANALYSIS OF THE COSTS OF LIVING WITH FLOODS

G1: ANOVA test

1.1. Hypothesis 0a: Is there a significant difference in the distribution of the current annual income between the socio-hydrological spaces?

To test the above hypothesis, we use a single-factor analysis of variance (ANOVA). We find that there exists a significant difference in current annual income between the socio-hydrological spaces (with $\alpha=0.05$).

Anova: Single Factor

SUMMARY

Groups	Count	Sum	Average	Variance
SH1	286	361806.3	1265.057	539348.3
SH2	298	289100	970.1342	345594.9
SH3	279	345125	1237.007	3666301

ANOVA

Source of Variation	SS	df	MS	F	P-value	F crit
Between Groups	15524131	2	7762066	5.233178	0.005508	3.006192
Within Groups	1.28E+09	860	1483241			
Total	1.29E+09	862				

From the table above we find that the average annual income in SH2 is USD 970, which is substantially lower than SH1 (USD 1265) and SH3 (USD 1237). Further, you see that the P-value is lower than the confidence level of 0.05.

Consequently, students' t-tests shows that SH2 significantly differ from both SH1 as SH3. SH1 and SH3 do not differ in annual income (with $\alpha=0.05$).

1.2. Hypothesis 0b: Is there a significant difference in the distribution of the current annual expenditure between the socio-hydrological spaces?

Anova: Single Factor

SUMMARY

Groups	Count	Sum	Average	Variance
SH1	286	370275	1294.668	1396803
SH2	298	286500	961.4094	323451
SH3	279	325575	1166.935	839192.8

ANOVA

Source of Variation	SS	df	MS	F	P-value	F crit
Between Groups	16546240	2	8273120	9.78059	6.31E-05	3.006192
Within Groups	7.27E+08	860	845871.3			
Total	7.44E+08	862				

Here we find that the average annual expenditure for SH2 is lowest (USD 962), followed by SH3 (USD 1167) and finally SH1 (USD 1295). Again, the P-value shows that there exists a significant difference in annual expenditure between the socio-hydrological spaces (with $\alpha=0.05$). Further, students t-test shows that SH2 significantly differ from both SH1 and SH3, while SH1 and SH3 do not.

1.3. Hypothesis 0c: Is there a significant difference in the distribution of the total wealth between the socio-hydrological spaces?

Anova: Single Factor

SUMMARY

Groups	Count	Sum	Average	Variance
SH1	286	9863089	34486.32	2.05E+09
SH2	298	1782068	5980.095	98264341
SH3	279	4923220	17645.95	6.4E+08

ANOVA

Source of Variation	SS	df	MS	F	P-value	F crit
Between Groups	1.2E+11	2	5.98E+10	64.99275	5.18E-27	3.006192
Within Groups	7.91E+11	860	9.2E+08			
Total	9.11E+11	862				

Again we find a significant difference between the spaces on total wealth. Additionally, compared to income and expenditure, we also see here a significant difference between SH1 and SH3.

G2: Chi-Square test: Remembering major floods by SHS

Observed Occurances	Socio-hydrological spaces			Observed Occurances	Socio-hydrological spaces		
	SHS1	SHS2	Total		SHS1	SHS3	Total
Flooded	34	55	89	Flooded	34	32	66
Non-flooded	21	0	21	Non-flooded	21	23	44
Total	55	55	110	Total	55	55	110

Expected Sources	Socio-hydrological spaces			Expected Sources	Socio-hydrological spaces		
	SHS1	SHS2	Total		SHS1	SHS3	Total
Flooded	44.50	44.50	89	Flooded	33.00	33.00	66
Non-flooded	10.50	10.50	21	Non-flooded	22.00	22.00	44
Total	55	55	110	Total	55	55	110

p =		3.49E-07	p =		0.69709

Observed	Socio-hydrological spaces		
Occurances	SHS2	SHS3	Total
Flooded	55	32	87
Non-flooded	0	23	23
Total	55	55	110

Expected	Socio-hydrological spaces		
Sources	SHS2	SHS3	Total
Flooded	43.50	43.50	87
Non-flooded	11.50	11.50	23
Total	55	55	110

p =		6.94E-08	

G3: Chi-Square test: Remembering major floods by socio-economic groups

p	Modpoor	Mod	Modrich	Rich
Poor	0.3130	0.6799	0.6799	0.3130
Modpoor		0.5499	0.5499	1.0000
Mod			1.0000	0.5499
Modrich				0.5499

Observed	Groups			Observed	Groups		
Occurances	Poor	Modpoor	Total	Occurances	Poor	Mod	Total
Flooded	39	34	73	Flooded	39	37	76
Non-flooded	16	21	37	Non-flooded	16	18	34
Total	55	55	110	Total	55	55	110

Expected	Groups			Expected	Groups		
Sources	Poor	Modpoor	Total	Sources	Poor	Mod	Total
Flooded	36.50	36.50	73	Flooded	38.00	38.00	76
Non-flooded	18.50	18.50	37	Non-flooded	17.00	17.00	34
Total	55	55	110	Total	55	55	110

p =		0.3130		p =		0.6799	

Observed Occurances	Groups Poor	Modrich	Total
Flooded	39	37	76
Non-flooded	16	18	34
Total	55	55	110

Expected Sources	Groups Poor	Modrich	Total
Flooded	38.00	38.00	76
Non-flooded	17.00	17.00	34
Total	55	55	110

p =   0.6799

Observed Occurances	Groups Poor	Rich	Total
Flooded	39	34	73
Non-flooded	16	21	37
Total	55	55	110

Expected Sources	Groups Poor	Rich	Total
Flooded	36.50	36.50	73
Non-flooded	18.50	18.50	37
Total	55	55	110

p =   0.3130

Observed Occurances	Groups Modpoor	Mod	Total
Flooded	34	37	71
Non-flooded	21	18	39
Total	55	55	110

Expected Sources	Groups Modpoor	Mod	Total
Flooded	35.50	35.50	71
Non-flooded	19.50	19.50	39
Total	55	55	110

p =   0.5499

Observed Occurances	Groups Modpoor	Modrich	Total
Flooded	34	37	71
Non-flooded	21	18	39
Total	55	55	110

Expected Sources	Groups Modpoor	Modrich	Total
Flooded	35.50	35.50	71
Non-flooded	19.50	19.50	39
Total	55	55	110

p =   0.5499

Observed Occurances	Groups Modpoor	Rich	Total
Flooded	34	34	68
Non-flooded	21	21	42
Total	55	55	110

Expected Sources	Groups Modpoor	Rich	Total
Flooded	34.00	34.00	68
Non-flooded	21.00	21.00	42
Total	55	55	110

p =   1.0000

186

Observed	Groups			Observed	Groups		
Occurances	Mod	Modrich	Total	Occurances	Mod	Rich	Total
Flooded	37	37	74	Flooded	37	34	71
Non-flooded	18	18	36	Non-flooded	18	21	39
Total	55	55	110	Total	55	55	110

Expected	Groups			Expected	Groups		
Sources	Mod	Modrich	Total	Sources	Mod	Rich	Total
Flooded	37.00	37.00	74	Flooded	35.50	35.50	71
Non-flooded	18.00	18.00	36	Non-flooded	19.50	19.50	39
Total	55	55	110	Total	55	55	110

p =			1.0000	p =			0.5499

Observed	Groups		
Occurance	Modrich	Rich	Total
Flooded	37	34	71
Non-flooc	18	21	39
Total	55	55	110

Expected	Groups		
Sources	Modrich	Rich	Total
Flooded	35.50	35.50	71
Non-flooc	19.50	19.50	39
Total	55	55	110

p =		0.5499

G4: Anova test for Relocation strategies by the respondents

Anova: Single Factor

SUMMARY

Groups	Count	Sum	Average	Variance
SHS1	286	71	0.248252	0.187278
SHS2	298	238	0.798658	0.161345
SHS3	279	116	0.415771	0.243779

ANOVA

Source of Variation	SS	df	MS	F	P-value	F crit
Between Groups	46.63684	2	23.31842	118.6167	3.2E-46	3.006192
Within Groups	169.0642	860	0.196586			
Total	215.701	862				

Anova: Single Factor

SUMMARY

Groups	Count	Sum	Average	Variance
Poor	192	146	0.760417	0.183137
Moderate poor	178	123	0.691011	0.214721
Moderate	169	48	0.284024	0.204565
Moderate rich	166	70	0.421687	0.245345
Rich	158	38	0.240506	0.183826

ANOVA

Source of Variation	SS	df	MS	F	P-value	F crit
Between Groups	39.00671	4	9.751677	47.35261	5.39E-36	2.382307
Within Groups	176.6943	858	0.205937			
Total	215.701	862				

G5: Anova test for change in occupation by the respondents

Anova: Single Factor

SUMMARY

Groups	Count	Sum	Average	Variance
SHS1	286	14	0.048951	0.046718
SHS2	298	21	0.07047	0.065724
SHS3	279	12	0.043011	0.041309

ANOVA

Source of Variation	SS	df	MS	F	P-value	F crit
Between Groups	0.121634	2	0.060817	1.180148	0.30773	3.006192
Within Groups	44.31869	860	0.051533			
Total	44.44032	862				

Anova: Single Factor

SUMMARY

Groups	Count	Sum	Average	Variance
Poor	192	13	0.067708	0.063454
Moderate poor	178	17	0.095506	0.086872
Moderate	169	4	0.023669	0.023246
Moderate rich	166	9	0.054217	0.051588
Rich	158	4	0.025316	0.024833

ANOVA

Source of Variation	SS	df	MS	F	P-value	F crit
Between Groups	0.62802	4	0.157005	3.074716	0.015749	2.382307
Within Groups	43.8123	858	0.051063			
Total	44.44032	862				

189

G6: Anova test for change in income by the respondents

Anova: Single Factor

SUMMARY

Groups	Count	Sum	Average	Variance
SHS1	286	276	0.965035	0.033861
SHS2	298	296	0.993289	0.006689
SHS3	279	271	0.971326	0.027952

ANOVA

Source of Variation	SS	df	MS	F	P-value	F crit
Between Groups	0.128964	2	0.064482	2.85738	0.057964	3.006192
Within Groups	19.40754	860	0.022567			
Total	19.5365	862				

Anova: Single Factor

SUMMARY

Groups	Count	Sum	Average	Variance
Poor	192	187	0.973958	0.025496
Moderate poor	178	174	0.977528	0.022091
Moderate	169	165	0.976331	0.023246
Moderate rich	166	161	0.96988	0.02939
Rich	158	156	0.987342	0.012578

ANOVA

Source of Variation	SS	df	MS	F	P-value	F crit
Between Groups	0.02719	4	0.006797	0.298947	0.8787	2.382307
Within Groups	19.50931	858	0.022738			
Total	19.5365	862				

G7: Anova test for change in expenses by the respondents

Anova: Single Factor

SUMMARY

Groups	Count	Sum	Average	Variance
SHS1	286	206	0.72028	0.202184
SHS2	298	236	0.791946	0.165322
SHS3	279	203	0.727599	0.198912

ANOVA

Source of Variation	SS	df	MS	F	P-value	F crit
Between Groups	0.911094	2	0.455547	2.418029	0.089703	3.006192
Within Groups	162.0205	860	0.188396			
Total	162.9316	862				

Anova: Single Factor

SUMMARY

Groups	Count	Sum	Average	Variance
Poor	192	151	0.786458	0.168821
Moderate poor	178	127	0.713483	0.20558
Moderate	169	124	0.733728	0.196534
Moderate rich	166	125	0.753012	0.187112
Rich	158	118	0.746835	0.190277

ANOVA

Source of Variation	SS	df	MS	F	P-value	F crit
Between Groups	0.534539	4	0.133635	0.706038	0.587904	2.382307
Within Groups	162.3971	858	0.189274			
Total	162.9316	862				

G8: Anova test for land loss by the respondents

Anova: Single Factor

SUMMARY

Groups	Count	Sum	Average	Variance
SHS1	286	126.5494	0.44248	3.389553
SHS2	143	218.0723	1.524981	5.986728
SHS3	279	241.2072	0.864542	17.67599

ANOVA

Source of Variation	SS	df	MS	F	P-value	F crit
Between Groups	112.346	2	56.17302	5.884341	0.002921	3.008498
Within Groups	6730.062	705	9.546187			
Total	6842.408	707				

Anova: Single Factor

SUMMARY

Groups	Count	Sum	Average	Variance
Poor	192	289.5054	1.507841	27.99725
Moderate poor	178	173.2543	0.973339	5.270592
Moderate	169	83.36738	0.493298	1.30381
Moderate rich	166	143.6656	0.865456	3.385565
Rich	158	96.21414	0.60895	4.385045

ANOVA

Source of Variation	SS	df	MS	F	P-value	F crit
Between Groups	113.3313	4	28.33282	3.138143	0.01414	2.382307
Within Groups	7746.48	858	9.028531			
Total	7859.811	862				

G9: Anova test for recovery from loss by the respondents

Anova: Single Factor

SUMMARY

Groups	Count	Sum	Average	Variance
SHS1	93	12	0.129032	0.113604
SHS2	265	16	0.060377	0.056947
SHS3	148	17	0.114865	0.102363

ANOVA

Source of Variation	SS	df	MS	F	P-value	F crit
Between Groups	0.465151	2	0.232576	2.886189	0.056713	3.013645
Within Groups	40.53287	503	0.080582			
Total	40.99802	505				

Anova: Single Factor

SUMMARY

Groups	Count	Sum	Average	Variance
Poor	152	6	0.039474	0.038167
Moderate poor	139	9	0.064748	0.060995
Moderate	71	8	0.112676	0.101408
Moderate rich	93	15	0.16129	0.136746
Rich	51	7	0.137255	0.120784

ANOVA

Source of Variation	SS	df	MS	F	P-value	F crit
Between Groups	1.099147	4	0.274787	3.450428	0.008529	2.389731
Within Groups	39.89888	501	0.079638			
Total	40.99802	505				

193

APPENDIX H: SUPPLEMENTARY ANALYSIS OF THE COSTS OF LIVING WITH FLOODS

H1: Major flooding events in the study area

The number of households were experienced flooding since 1960s are shown in Figure H1a. According to our household surveys; total 33 major floods were observed in our study area since 1960s. Flood affected area of Bangladesh is shown in Figure H1b. We correlated between flood memories of the households with total flood affected area of Bangladesh and as expected we did not find any good correlation between them (Figure H1c). It is because percentages of study area inundation is different than the percentages of area inundation of the whole country. We also analyzed the flood memory with inundated area of our study area for the years from 1998 to 2004 and it shows a very good correlation between them (Figure H1d). It means the memory of the HH respondents is good enough to remember flood events.

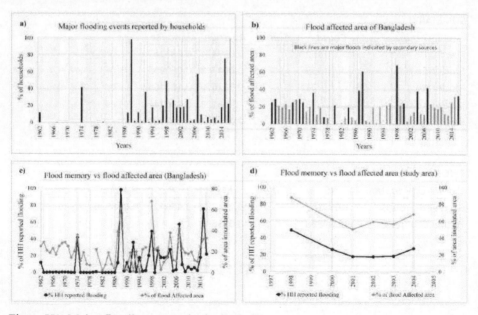

Figure H1: Major flooding events in the study area.

H2: Correlation between flood severities with the reported losses

To do this analysis we have used time series household data from 1960 to 2016 on flooding events and their respective losses. We correlated the damage data with yearly maximum water level of the Jamuna River but could not find any good correlation between them (Figure H2a). Then we analyzed the damage data with flood depth at homestead and agricultural land but here we also could not find any good correlation between them. We analyzed the flood damages with the inundated area of the study area and observed similar trends but the correlation was not so good (Figure H2b).

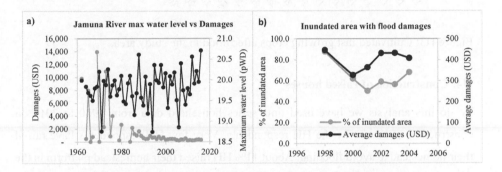

Figure H2: Correlation between flood severities with the reported losses.

H3: Changing cropping pattern

To analyses the change in cropping pattern due to flooding event we have used the time series data on cropping pattern from 1960 to 2015. Most of the households do not change their cropping pattern due to flooding. With their indigenous knowledge, they have developed a unique technique to cultivate rice and they are following this technique in every year. Still due to abnormal floods, this technique does not work and they lose everything. About 20% HH cultivated fast growing crops after the abnormal flood event in 1988 and about 15% HH cultivated fast growing crops after the flood event in 2007 and 2015 (Figure H3). Few farmers were found to put fallow their agricultural land during flood season as they do not want to lose their investments in the field.

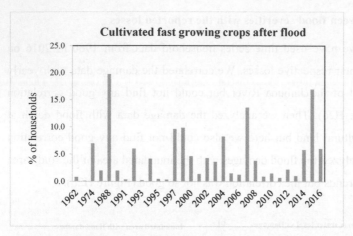

Figure H3: Cultivated fast growing crops after flood in the study area.

H4: Construction of raised houses

To do this analysis we have used time series household data from 1960 to 2016 on construction of houses by the HH respondents. Very few HH respondents have raised their homestead to avoid flooding. About 11% HH raised their homestead platform in the study area (Figure H4a). More households from rich and moderately rich households have raised their homestead platforms (Figure H4b).

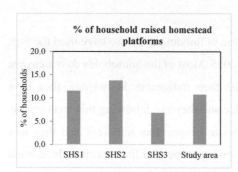
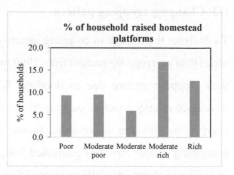

Figure H4: Percentages of households raised homestead platform in the study area.

H5: Relocation distances of households

The respondents relocated within 5 km from their previous locations (Figure H5a), and many of the poor and moderate poor household are thinking to relocate again to save from riverbank erosion (Figure H5b).

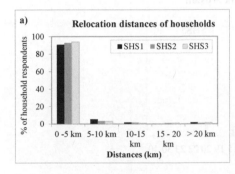
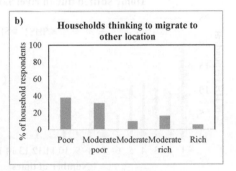

Figure H5: Relocation distances of households in the study area.

H6: Total assets loss

Our survey data includes time series on land loss, agricultural crop loss, homestead loss and other assets loss due to flood and riverbank erosion. If we look at the total assets loss, households in SHS3 have lost most assets, at around 2,000 USD per year (Figure H6a). By socio-economic status, poor households have lost most assets at more than 2,000 USD (Figure H6b).

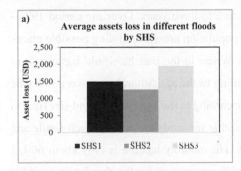
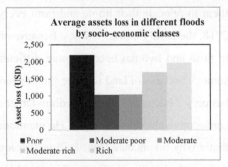

Figure H6: Average assets loss by households in different floods in the study area.

H7: People moved due to river bank erosion

Households in the study area are moving their homes due to riverbank erosion. It found that one household had to move 40 times in their entire life (Figure H7).

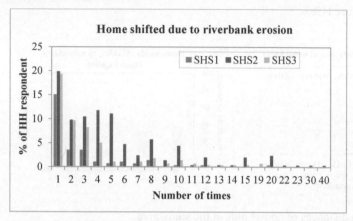

Figure H7: Households shifted their homes due to river bank erosion.

H8: Story of Mr. Mohammad Abdul Mazid, an example of living with floods in the study area

Mr. Mazid, who is 81 years of age, living in a char of Jamuna River in Fazlupur union of Fulchari upazila of Gaibandha district. He is an agricultural farmer. He organize his livelihoods around the river. He for instance grow rice, maize, jute and vegetables on a field by the Jamuna River char. The ever-changing course of the river or a flood forces them to relocate their home and farm every two to three years. Over the period 1973 - 2018, the family moved 9 times. Because of population pressure, finding possible places to farm and live has become more difficult. Where in the past he would has access to some 5 hectares of land (a large famer according to the agricultural land size), today he has only 2 hectres of land (a medium famer according to the agricultural land size) due to consecutive riverbank erosion. He mentioned that his lands are not so much fertile and that's why his income from land is also low. His monthly income is only about 60 US dollars. As his income from agricultural land is not enough for the family, he limit his daily needs. His house is made of locally available materials with earthen floor, wood with tin on the top of the house. Mr. Mazid is an example of many people living in the

Jamuna floodplain and its char. Many people like him used to be rich earlier but due to flooding and erosion they have become poor. His timeline of moving from one place to another place due to river bank erosion is presented in the following Figure H8.

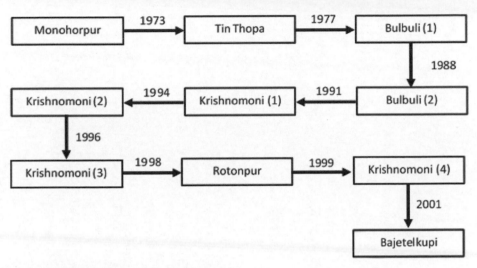

Figure H8: Moving timeline of Mr. Mazid

ABOUT THE AUTHOR

Md Ruknul Ferdous was born in Dinajpur, Bangladesh in 1980. He studied Water Resources Engineering at the Bangladesh University of Engineering and Technology (BUET), Dhaka, Bangladesh, where he graduated in 2005. He completed a Master of Science in Water Science and Engineering (Specialization: Hydrology and Water Resources) at IHE Delft Institute for Water Education, Delft, the Netherlands in 2014.

From 2005 to 2014, M. R. Ferdous worked at CEGIS, a public trust under the Ministry of Water Resources, Bangladesh as a Water Resources Expert and River Morphologist. In this capacity, he became involved in research that combined insights from hydrology, river morphology, water management, climate change and environmental studies. He worked in a number of projects in which he used such insights to help develop flood forecasting systems, produce vulnerability assessments, develop ways to adapt to natural hazards and climate change. He also engaged in environmental and social impact assessment of flood management projects in Bangladesh.

In 2014, he started his PhD research at IHE Delft Institute for Water Education, Delft, the Netherlands and the University of Amsterdam, the Netherlands under the program "Hydro-Social Deltas: Understanding flows of water and people to improve policies and strategies for disaster risk reduction and sustainable development of delta areas in the Netherlands and Bangladesh". His research interests are human-water interactions, hydrology, flood and river bank erosion, and river morphology.

LIST OF PUBLICATIONS

- **Ferdous, M.R.**; Wesselink, A.; Brandimarte, L.; Slager, K.; Zwarteveen, M. and Di Baldassarre, G. (2019). The Costs of Living with Floods in the Jamuna Floodplain in Bangladesh. Water, 11, 1238, https://www.mdpi.com/2073-4441/11/6/1238.

- **Ferdous, M. R.**; Wesselink, A.; Brandimarte, L.; Di Baldassarre, G. and Rahman, M. M. (2019). The levee effect along the Jamuna River in Bangladesh. Water International, https://doi.org/10.1080/02508060.2019.1619048.

- **Ferdous, M.R.**; Di Baldassarre, G.; Brandimarte, L.; Wesselink, A. (2019). The interplay between structural flood protection, population density and flood mortality along the Jamuna River, Bangladesh. Manuscript is accepted for publication in the Journal Regional Environmental Change.

- **Ferdous, M. R.**; Wesselink, A.; Brandimarte, L.; Slager, K.; Zwarteveen, M. and Di Baldassarre, G. (2018). Socio-hydrological spaces in the Jamuna River floodplain in Bangladesh. Hydrol. Earth Syst. Sci., 22, 5159-5173, https://doi.org/10.5194/hess-22-5159-2018.

- Di Baldassarre, G.; Yan, K.; **Ferdous, M. R.** and Brandimarte, L. (2014). The interplay between human population dynamics and flooding in Bangladesh: a spatial analysis. Proceedings of the International Association of Hydrological Sciences, 364, 188-191.

- Sarker, M. H.; Thorne, C. R.; Aktar, M. N. and **Ferdous, M. R.** (2014). Morpho-dynamics of the Brahmaputra-Jamuna River, Bangladesh. Geomorphology, 215, 45-59.

- Sarker, M. H.; Akter, J.; **Ferdous, M. R.** and Noor, F. (2011). Sediment dispersal processes and management in coping with climate change in the Meghna Estuary, Bangladesh. IAHS-AISH publication, 203-217.

- Mondal, M. S.; Chowdhury, J. U. and **Ferdous, M. R.** (2010). Risk-based evaluation for meeting future water demand of the Brahmaputra floodplain within Bangladesh. Water resources management, 24(5), 853-869.

Conference proceedings

- **Ferdous, M. R.**, Wesselink, A., Brandimarte, L., Slager, K., Mynett, A., & Zwarteveen, M. (2017). Living with floods in the Jamuna floodplain (Bangladesh): fight or flight? Technological and societal responses. In EGU General Assembly Conference Abstracts (Vol. 19, p. 513).

- **Ferdous, M.R.**, Brandimarte, L., Di. Baldassarre, G., (2015). Understanding the dynamics interplay between hydrological and social processes of southwest coastal region of Bangladesh. Abstract presented in 26th IUGG General Assembly of the International Union of Geodesy and Geophysics (IUGG) held in Prague, Czech Republic, June 22 –July 2, 2015.

- Hore, S.K.; Sarker, M.H.; **Ferdous, M.R.**; Ahsan, M. and Hasan, M.I., (2013). Study of the off-take dynamics for restoring the Gorai River. 4th International Conference on Water & Flood Management (ICWFM-2013), Dhaka, Bangladesh.

- Sarker, M.H.; Akter, J. and **Ferdous, M.R.**, (2011). River bank protection measures in the Brahmaputra-Jamuna River: Bangladesh Experience. International Seminar on River, Society and Sustainable Development at Dibrugarh University, India.

- Sarker, M.H.; Akter, J.; **Ferdous, M.R.**, Noor, F., (2011). Sediment dispersal processes and management in coping with climate change in the Meghna Estuary, Bangladesh. Proceedings of the Workshop on Sediment Problems and Sediment Management in Asian River Basins, Publication: 349, pp. 203-218, Hyderabad, India.

- Shah, M.A.R.; **Ferdous, M.R.** and Hassan, A., (2007). Potential Threats of Power Sector on Water Resources. International Conference on Water and Flood Management at Institute of Water and Flood Management, Bangladesh University of Engineering and Technology (BUET), Dhaka. Bangladesh.

- Bhuiyan, F.; **Ferdous, M.R.**; Haider, R. and Wormleaton, P. (2005). 3-Dimensional flow structure and Scour in Meandering Channel with Floodplain. Proceedings 3rd International Conference on Civil Engineering, March 9- 11, 2005, IEB, Dhaka, pp.15-22.

T - #0091 - 071024 - C228 - 240/170/12 - PB - 9780367902131 - Gloss Lamination